AF540127

Rivers, Dams and Development

Rivers, Dams and Development

Issues and Dilemmas

Edited by

Prabha Shastri Ranade

2010

Icfai Books
The Icfai University Press

RIVERS, DAMS AND DEVELOPMENT: ISSUES AND DILEMMAS

Editor: Prabha Shastri Ranade

First Edition: 2010
Printed in India

Published by

The Icfai University Press
52, Nagarjuna Hills, Punjagutta
Hyderabad, India – 500 082
Phone: (+91) (040) 23430–368, 369, 370, 372, 373, 374
Fax: (+91) (040) 23352521, 23435386
E-mail: info@icfaibooks.com, icfaibooks@icfai.org, ssd@icfai.org

ISBN: 9788131408421

Editorial Team: Shashikala Choudhary and Neelima A
Quality Support: K Prabhakar and Ch Ramesh

Contents

Overview

Rivers have enormous value to society as a major source of surface water resources. Ancient and modern communities alike have depended on rivers for livelihood, commerce, habitat and the sustaining ecological functions they provide. Water as a natural resource has a tremendous effect on every aspect of development. Human demand for water and water-related services is increasing significantly due to the growing population and the rising level of economic activity. By 2025, one-third of the population of the developing world is estimated to face severe water shortages. Yet, even in many water-scarce regions, large amounts of water annually flow out into the sea.

During the 20th century, large dams emerged as one of the most significant and visible tools for the management of rivers. More than 45,000 large dams around the world have played an important role in helping communities and economies harness water resources for food production, energy generation, flood control and domestic use. Current estimates suggest that around 30-40% of irrigated land worldwide now relies on dams and dams generate 19% of world electricity.

Official statistics suggest that large dams supply approximately 30-35% of irrigation water in the two countries with the largest irrigated areas – India and China.

Dams also alter and divert river flows, affect existing rights and access to water, and create significant impact on livelihood and environment. Global estimates of the magnitude of impact say that 60% of the world's rivers have been affected by dams and diversions. Also, around 40-80 million people have been displaced by dams. With increasing water demand and greater awareness of environmental concerns and social responsibilities, the complexity of river management, dam planning and operation has increased greatly in recent years. It is no longer acceptable to optimize dam operation on the basis of economic criterion alone. Participation of stakeholders in the decision-making process and increased equity in the distribution of benefits are prerequisites to mitigating dam-related conflicts and ensuring sustainable projects. Appropriate use of decision support systems can enhance economic, social and environmental benefits.

This book aims at exploring the significance of rivers and dams from economic, ecological, social and cultural angles. It attempts to present general patterns and trends in the management of rivers as well as planning and operation of large dams. It analyzes the central issues in the large dams debate – the benefits and adverse impacts – and the positions taken by different groups of planners, public policy-makers and various stakeholders. This book is divided into three sections: "Rivers and Dams: An Overview", "Key Issues" and "Country Experiences".

Section I: Rivers and Dams: An Overview

The first section of the book introduces the importance of rivers in developing culture, civilization and economy. Various concepts and approaches in river management and dam management, riverfront projects and inter-linking of rivers are covered. Usefulness of dams in managing rivers, irrigation, hydroelectric power generation, flood management, etc., are discussed in this section.

The first article is "**Dams and River Management for Development**" by *Prabha Shastri Ranade*. The article presents dams as a tool for managing rivers in the 20th century. Thousands of large dams around the world help in harnessing river water for food production, energy generation and flood control. It describes the types of floods, their causes and their impact. It also describes the dams' boon as they aid employment generation and economic development and their bane as they cause severe environmental problems.

The next article "**Who takes Decision for Large Dams? Many Questions, Few Answers**" by *Himanshu Thakkar,* shows the real picture of political economy of dam construction in a nutshell. It says that the Government of India as well as the World Bank are supporting large dams and also other projects like interlinking of rivers. The author strongly feels that the public should be included in decision-making process involved in building dams.

The third article is "**Interlinking of Rivers in India—Assessing the Justifications**" by *Jayanta Bandyopadhyay* and *Shama Perveen*. They present the proposal of river interlinking in India, made by the National Water Development Agency. The article critically examines the assumptions behind, and the main justifications extended for, the project. It views this interlinking project, which is considered a must by many politicians, as an extremely cost-ineffective way of improving irrigation process in India. The latest information on this interlinking project is that agreements have been made among the states concerned for two of the peninsular links.

Article four "**Irrigation within Integrated Catchment Management**" by *Geoff McLeod* and *Scott Keyworth,* presents the investments in irrigation projects supported by Australia's Murray- Darling Basin Commission since 1992. The focus of these investments is to develop strategies for sustainable irrigation practices within an Integrated Catchment Management context. The article says that the clarity of vision and proper planning and information management are required to achieve the desired outcomes.

Section II: Key Issues

The second section examines the management and economic performance of dams, taking into consideration the huge investments made in creating dam infrastructure. Environmental performance of dams and their impact on river ecology are assessed. The section also covers social performance of dams and the issue of displacement of people.

The first article of this section **"To Dam or not to Dam? Five Years on from the World Commission on Dams"** is a report prepared by WWF, which analyzes the impact of recommendations for dam construction in WCD's report launched five years ago. It includes six case studies of different countries where the government concerned has failed to comply with the WCD recommendations. WWF has found a positive effect of the recommendations as it is slowly changing the viewpoint regarding dams.

The next article is **"Managing the Environmental Impact of Dams"** by *Matthew P McCartney* and *Hilmy Sally*, who have discussed the pros and cons of dams. It says that, often the lack of foresight and planning for dams in isolation from catchment affects the mankind and the environment harmfully. It provides a brief review of the consequences of ecosystems and biodiversity resulting directly from the presence of dams on rivers. Strategies to protect the environment are described.

The article **"Narmada and the Myth of Rehabilitation"** by *Mike Levien* is based on the report of Shunglu Committee which was appointed to verify the families affected by Sardar Sarovar Project. The author says that the committee has ignored the Supreme Court rulings on the issue and has given a positive report, though there is a clear recognition of the incomplete process of rehabilitation.

The next article is **"Spotlight on Indus River Diplomacy: India, Pakistan and Baglihar Dam Dispute"** by *Robert G Wirsing* and *Christopher Jasparro*. This article talks about the river water issues and the severity of rivalry between nations. It mentions the dispute over the Baglihar dam on Chenab river in Jammu and Kashmir. The Pakistanis have raised the issue of the design of Baglihar dam because, according to Indus Water Treaty (IWT), Pakistan has the ownership of Chenab river water. The author says that this issue can intensify the already existing tension between Pakistan and India. To solve this issue, the World Bank gave a verdict recently and both the countries are ready to abide by it.

The last article of this section is **"Muted Elation"** by *T S Subramanian.* This article is on the matter of final award declared by Cauvery Water Disputes Tribunal on the 5th Feb, 2007 regarding water-sharing among Tamil Nadu, Karnataka, Kerala and Puducherry. This brought an end to the issue which caused more than 40 years of unrest in these states. The author says that there was a sense of elation and justice in Tamil Nadu. But it all depends on whether Karnataka will follow the verdict guidelines.

Section III: Country Experiences

The tenth article of the book **"Ecology, Planning, and River Management in the United States: Some Historical Reflections"** by *Martin Reuss* is an example of river management in the US. The author says that the governments justify river management projects only on the economic criterion, ignoring the intangible benefits. Large dam projects force basin inhabitants from their homes, and chemical and nuclear pollutants threaten the environment. The author says that now the integrated water resources management, sustainable development and river basin management are becoming more dominating in the US. He believes that any planning regarding water should include ecologists who can give proper advice.

Article eleven is **"The Three Gorges Project and Flood Control of the Yangtze River"** by *Qihui Yu, Yongjun Zhang* and *Weiyu Tang.* The Three Gorges Project is one of the world's largest dams. The authors say that flood control is the most important task of the regulation and development of the Yangtze river basin. They analyse the role of TGP in flood control on Yangtze river. TGP is a multi-purpose hydro-development project producing comprehensive benefits mainly in flood control, power generation and navigation improvement.

In this article **"Ethiopia's Water Dilemma"** the author *Uwe Hoering*, presents the situation of unpredictable distribution of water that affects the economy of Ethiopia. It has an enormous water potential but still untapped. The article includes the Nile river basin initiative taken by the World Bank. It also says that the World Bank and many other multilateral financial institutions are ready to lend billions of dollars to develop the water infrastructure in Ethiopia. Though large dams can be helpful in combating its poverty, developmental analysts say that small-scale water projects can cater to the immediate needs of the rural poor.

The next article **"Flooding the Future: Hydropower and Cultural Survival in the Salween River Basin"** is prepared by EarthRights International. This article presents the decision of Thai Government, SPDC and military regime of Burma to build a series of dams on Salween River. EarthRights International has conducted fact-finding study of such development projects in Burma since 1996 and found that the dams would result in human rights abuses and environmental degradation. This article summarizes the findings of several recent studies on this highly contentious issue, and offers a series of recommendations for addressing them.

The article **"Kalpasar: India's Most Ambitious Project"** by *Prabha Shastri Ranade* talks about the Kalpasar project which is one of the key projects of Gujarat. A 64-km dam is proposed to be built across gulf of Khambhat from Ghogar in Bhavnagar district to Hansot in Bharuch district. The idea is to trap water from 12 rivers of Gujarat to construct a freshwater lake. Through a 660-km canal system, 1.05 million hectares of land in coastal Saurashtra will be irrigated. It will produce power to the tune of 6,000 mw through underwater turbines in the sea. A multi-lane highway and a railway line have also been planned across the length of the dam.

The last article is **"Behind Bhakra, Beyond Bhakra"** by *Shripad Dharmadhikary*. It is the concluding part of a report "Unravelling Bhakra: Assessing the Temple of Resurgent India". It highlights the effects of Bhakra project which was the first of the storage dams in the Indus Basin. The author has mentioned the pros and cons of diversion and storage systems. He has clarified many misconceptions regarding Bhakra by comparing them with his findings. He has also presented some serious issues like diminishing resources and increasing disequilibrium in the surrounding area.

Section I

Rivers and Dams: An Overview

1

Dams and River Management for Development

Prabha Shastri Ranade

Dams have emerged as one of the most popular tools for the management of rivers in the 20th century. There are over 45,000 large dams around the world to harness river water for food production, energy generation and flood control. The article describes the types of floods, their causes and their impact. Floods have taken a large toll of human lives over the years and caused large economic losses. This article also describes dams' bane as they cause severe environmental problems and boon as they also serve as a good tourist attraction.

Introduction

As India receives seasonal and variable rainfall, it needs to store water to meet its requirements throughout the year. Water deficit is the difference between the demand for water and available supplies. India's water deficit is the largest of any country in the world. This has caused Indians to turn to dams to store water.[1] A river is a large body of flowing water constrained in a channel. Geologically it is the main trunk of the drainage system. The discharge volume of an individual stream/river is changing from month to month and year to year. It is estimated that rivers generally overflow their banks (flood) at least once in every two years. However, the recurrence interval (time between floods) varies from river basin to river basin and also varies with the size of flood. When the rivers overflow their banks, it normally results in flood disasters. Flood is the overflowing of water onto land that is normally dry.[2] Flood is a high water stage in which river water overflows its natural or artificial banks over the adjoining land, which is usually dry. It's a temporary condition of inundation. It involves an unusual and rapid accumulation or runoff of surface waters from any source. A condition of floods exists when the discharge of a river cannot be accommodated within the margins of its normal channel. Then the water spreads over the surrounding region, irrespective of whether it is the crop land or forest land.

Flood plain is a belt of low flat ground bordering the channel on one or both sides inundated by stream waters about once a year. In India monsoon rain brings abundant supplies of surface water, which results in increasing the water table. There is also ample soil moisture. So with more rainfall there is more runoff than can stay within the river channel. Such annual inundation or overflow of water from the river is considered as flood. Floods can be measured for height, peak discharge, area inundated and volume of flow. For practical purposes, a particular stage of gauge height at a given place is designated as the flood stage, a critical level above which over bank flooding may be expected to set in. These factors are important to judicious land use, construction of bridges and dams and prediction and control of floods.

The Antiquity of Floods

The history of human civilization is closely linked to the world's rivers. Society has developed an intimate relationship with rivers. Floods are not new to the people and they have learnt to live with them. Floods are a phenomenon older than the cities that exist on the banks of the rivers. During the 15th and 17th centuries most people believed that land features were either formed when the earth was created, or were the result of sudden violent catastrophes, such as floods, which were usually caused by supernatural forces. The flood theorists believed that flood was responsible for a great many features of the landscape. Where the topography has been carved out by glaciers or deeply eroded by rivers, the sea has flooded the lower end of these valleys. Almost all coasts of the world have been flooded in the past and it was probably this effect that led to the innumerable local legends of catastrophic floods submerging the lands of prehistoric people. Great flood of the Bible is one such best known of the legend. Ancient Indian literature is full of references to these devastating floods. Floods have been occurring in several countries since time immemorial. Major Himalayan rivers behave as erratically as small mountain torrents. Floods are not new to the Indo-Gangetic plains. During the 3,500 years of recorded human settlements in the Ganga basin, there have been many floods of gigantic proportions. Even the state of Rajasthan, which is considered as a drough-prone area experiences floods. A study reveals that Rajasthan is not only a drought prone area, but also has some flood -prone areas. Twenty districts in Rajasthan develop flood conditions each year in different areas with different magnitudes.[3]

Types of Flood

Floods are the most common, climate-related disasters. The following types of floods are commonly experienced: seasonal flooding, flash flooding and urban flooding due to inadequate drainage facilities and unplanned expansion. Summer floods are produced directly by torrential rains from invading moist tropical air masses. Natural drainage gets blocked or obstructed by human activities and encroachment of the flood plains. Urbanization and highway construction are major sources of excessive sediment causing channel aggravations. Drainage congestion becomes a serious problem with the runoff from the fields accumulated near the embankments. July 2005 floods in Mumbai were the result of natural drainage getting choked or obstructed by human activities due to illegal construction. Bridges, roads and railway lines built against natural drainage compound the problem. Flash flood is a sudden unexpected torrent of muddy and turbulent water rushing down a canyon. A flash flood can take place in a single tributary, while the rest of the drainage basin remains dry. Rajasthan's Barmer district experienced such floods in September 2006. The suddenness of its occurrence causes a flash flood to be extremely dangerous. Flash flood waters can move at a very fast speed and can roll boulders, tear out trees, destroy buildings, and obliterate bridges. Walls of water can reach heights of 10 to 20 ft. and are accompanied by a deadly cargo of debris.

Causes of Flood

Inability of rivers to cope with increased discharge upstream is the cause of floods. Although nature is generally blamed, many times it is the direct or indirect effect of human action that triggers a flood.[4] The growing world population and people's demands have put more pressure on rivers. Intense local rainfall can trigger severe local flooding. 500 mm rainfall in 24 hours makes adequate drainage on flood land impossible. This is responsible for devastating floods in themselves, or may add to flooding from other sources. Floods also result due to upstream dam failure and from tsunamis, the mountainous sea waves caused by earthquakes. In many parts of the world, spring is the time of floods because the flow of rivers is augmented not only by spring rains but also by melting of snow.

Some disastrous floods result from ice jams during the spring rise from storm tides. Fluvio-glacial landscape is a region of catastrophic flooding by glacial melt waters. South Asia is more prone to floods. In India Himalayan rivers like Ganga and Brahmaputra are the carriers of enormous quantity of water. Assam is the most flood-prone state in India. Other flood-prone states in India are Uttar Pradesh, Bihar, West Bengal, Assam and Orissa. Natural factors contribute more to floods in Assam. Landslides often block Himalayan rivers. When these landslide dams burst, they cause a flood pulse, which triggers off more landslides. At least one-third of Bangladesh is flooded every year. Bangladesh is particularly susceptible to floods emerging from the sea. As many as 80 million people in Bangladesh are vulnerable to floods each year. Summer rains wreck havoc across the flood-prone Yangtze practically every year and low lying plains in Central China are affected.

In 1973, Pakistan's Punjab and Sind provinces experienced floods, which resulted from abnormally heavy rainfall in lower regions of Indus system. It also coincided with snow melt and monsoon rain in the upper reaches of the catchments area. Sri Lanka's catastrophic flooding in 1957 resulted from excessive (four times the average) rainfall associated with intense cyclonic depression. The heavy downpour in Barmer district of Rajasthan in September 2006, led to flash floods in desert. If suitable rainwater harvesting structures are developed, it could harvest and conserve 90-145 million cubic metres of runoff a year in addition to recharging aquifers. The recent flood indicates a solution to a potential crisis through groundwater recharge.[5] In 2000 and 2001 Mozambique was hit by floods, which affected about one-fourth of its population and destroyed much of its infrastructure. In 2002 a severe drought hit the country including previously flood stricken areas. In February 2007 the Zambezi river broke its banks, flooding surrounding areas in Mozambique.[6]

Floods by Tidal Currents

As the tide begins to rise, a landward current, the flood current begins to flow and gain strength to attain a maximum speed at about mid-tide. The ebb current is stronger than the flood current. The rivers contribute a considerable discharge of fresh water from the land, which must escape to the sea. This stream discharge augments the ebb current, but opposes the flood current. The rise of a river stage to its maximum height or crest, followed by a gradual lowering of stage is termed the flood wave. Floods are associated with tidal events induced by typhoons in coastal areas. Storm surge associated with cyclone may raise sea level temporarily by up to 3 metres. Eastern coast of India is badly hit by cyclonic flood waves every year.

Floods Resulting from Excess Water Released Through Dams

Environmentalists criticize that dams have become the main cause of floods instead of providing protection against flood. There are several examples of floods, which resulted after the release of excess water in their reservoirs. Heavy rainfall across Zimbabwe, Zambia and Malawi in February 2007 added into the reservoir of Mozambique's main hydro-electric dam, the Cahora Bassa and filled it to capacity. Flood gates needed to be opened to prevent the dam walls from bursting. Excess water released from the dam resulted in disastrous floods. Floods in Mozambique have left 68,000 people homeless.[7] Surat, the coastal city of Gujarat located at the junction where Tapi river meets the Arabian Sea, experienced a devastating flood in the monsoon season of 2006. Torrential rainfall in the catchment area resulted in filling up of all the reservoirs/dams. In order to prevent the dams from overflowing and spreading of the water in the surrounding region, the sluice-gates of all the major dams were opened, and the Tapi river, unable to drain the excess water to the Arabian sea due to various constraints at the mouth, overflowed within the Surat city, resulting in severe damage to life and property, including the industrial establishments. During the last week of August 2000 Andhra Pradesh received the highest rainfall in the past 46 years. Hyderabad and

Secunderabad were inundated by overflow from the neighboring lakes, forcing thousands of people to take refuge on the roofs of their houses or to flee the city. The excess flow of water from Hussain Sagar due to lifting of gates was partly responsible for inundation of several parts of Hyderabad.[8]

Impact of Floods

The life-sustaining waters of a river also have the ability to destroy us completely as floods.[9] Floods are important to man because of their associated impacts and consequences. The effects of floods on human lives range from unqualified blessings to catastrophe. Flood waters can be extremely dangerous. The force of 6 inches of swiftly moving water can knock people off their feet. Just 2 feet of moving water has the power to move away standing cars. Floods have taken a large toll of human life over the years and also caused large economic losses. It is estimated that flood deaths in India account for 20 percent or one-fifth of the global death count. Floods cause damage to crops, houses and public property. In case of Assam the major factor responsible for the underdeveloped state of agriculture is the damaging impact of disastrous floods.[10] This study shows that floods have been the major cause of lower level of development of Assam.

Floods as a Boon

Floods, though classed as a natural disaster, are not entirely a bad phenomenon. They also bring several ecological advantages. The regular, seasonal spring floods of the Nile river prior to construction of the Aswan dam were much awaited to provide moisture for the fertile flood plains of its delta. Floods bring with them annually the deposition of the rich silts. In the Ganges and the Brahmaputra river basins, the intensive rice cultivation is possible because of the huge deposition of silt brought by the flood waters for years together. The flood waters from the flash floods in 2006 monsoon season, created small lakes in certain regions of Rajasthan. The water of such reservoirs was used for agriculture. The floods indicate a solution to a potential crisis through groundwater recharge.[11] Heavy rainfall once in four to five years is a common phenomenon in many areas. Dams, tanks, anicuts can be built along the seasonal rivers which have flash floods.

Flood Damage in India

Property damage from flooding costs over billions of dollars every year. It has been estimated that annual flood losses in some countries are much more now than what they were before. Floods are a major cause of human misery in India. In India annual flood damage has increased several times. It was estimated as an average of Rs.60 crore a year during the 1950s. It exceeded Rs.2307 crore a year during the 1980s. About 40 million hectares of land is at risk from flooding each year. The flood-affected areas shot up from an average of 6.4 million hectares a year in the 1950s to 9 million hectares a year in over 30 years time. The annual flood-affected population in India has risen faster than India's population. According to the Indian Government, one out of every 20 people in the nation is vulnerable to flooding.

Various District Gazetteers published during the British period described the havoc caused by the large movements of the rivers from place to place. The Kosi river has moved 120 km westward in the past 250 years through 12 distinct channels. Floods bring untold miseries for the people living in these areas where large tracks are submerged every year. During a year of catastrophic flood, some 20 mm deep soil of the Himalayas gets eroded. This is a very high erosion rate. As siltation takes place within the embanked area, the riverbed rises rapidly. Dredging of the mighty rivers has proved impractical. Some studies have pointed that though afforestation will help the local economy, it will not prevent floods in Himalayan rivers.

Floods, which are the natural disasters, turn out to be social disaster as well. Floods affect the poor more as they occupy the flood-prone areas. Survival after the floods is always a problem, as water remains in the field for a considerable time. The vast area gives the look of an ocean dotted by submerged villages. No place is to be found even to airdrop food packets. The government is not able to give sufficiently early warning to several areas where floods occur. According to officials about 1600 families of 16 villages were severely hit by the floods in Barmer district (Rajasthan) in August 2006. Over 100 people lost their lives in these floods, about 80,000 people were affected and over 45,000 cattle perished.[12] The fury and extent of floods in Assam have increased manifold over the years. In 1995, 55.99 lakh people spread over 7008 villages were affected by floods. In

1998, 9.7 lakh hectares of land and 47.10 lakh people were affected.[13] The floods of river Krishna, the main river of the state, affect Andhra Pradesh regularly. The flood of 1996 destroyed 714,040 dwellings and killed 1,685 people and 66,694 cattle. Gujarat by and large is drought prone. Nevertheless floods in the Narmada and Sabarmati rivers do occur. Floods in 1996 destroyed 54,575 dwellings and killed 117 people.[14]

Flood Abatement Measures

The art of flood control was well-known in ancient India. The best protection during a flood is to leave the area and take shelter on a higher ground. Throughout history, people have sought to tame and control rivers by constructing dykes, diversions and dams. Improvements of channels, construction of protective levees, and storage reservoirs, soil and forest conservation to retard and absorb runoff from storms are some of the measures. Government flood control measures mainly consist of construction of dams and embankments. Efforts are made to modify the lower reaches of the river where flood plain inundation is expected. To cope with repeated disastrous floods, vast sums of money have been spent on a wide variety of measures to reduce flood hazard, to detain and delay runoff by various means on the ground surface and in smaller tributaries of the watershed.

In India over 400 km of embankments have been built annually since 1954. It has been observed that embankments have disrupted the natural drainage system in the flood plains. Recurrence of embankment failures has assumed unmanageable proportions. An alarming increase in flood-affected area, despite protection makes it obvious that flood control measures have failed to deliver the goods. Flood insurance programs are introduced in the United States and other advanced countries. The insurance companies claim that buying flood insurance is the best thing you can do to protect one's home, business, family and financial security. After the tsunami disaster, the idea of insurance against natural disasters, including floods, is slowly picking up in India as well.

Constructing Dams for Managing Floods

Dams are the largest structures built by the humankind. In recent years dam construction has become a controversial issue. The idea of controlling floods through dams and embankments is relatively recent, in sharp contradiction to the idea of living with floods, practiced in the past in several countries. There are now more than 45,000 large dams across the world. The best dam sites have been used.[15] The WWF report on free flowing rivers points out that of 177 of the world's large rivers only one-third remain free flowing.[16] The earliest recorded dam is believed to have been on the Nile river at Kosheish, where a 15m high masonry structure was built around 2900 B.C. to supply water to capital of Memphis.[17] During the British period the British government sent Sir Arthor Cotton, an engineer, on special appointment for managing irrigation system in India. Godavari Anicut, now called as Dowleswaram barrage was constructed by Sir Arthor Cotton in Godavari district for managing the floods of Godavari river and irrigating the rice fields. Barrage construction was started in 1843 and was completed in 1852. Anicut is a dam made in the course of a stream for the purpose of regulating the flow of a system of irrigation. Hoover dam in North America is the first dam of the modern era. Hoover dam completed in 1935 symbolised man's conquest over nature and became a model for other countries. Thereafter, about 140 countries built dams by spending over $2 trillion. Dams are now found over 60 percentage of the world's 200-plus major river basins. Over 400 large dams have been constructed in India. Their turbines generate a fifth of the world's electricity supply and the water they store is utilised for producing one-sixth of the earth's food production.[18]

Impact of Dams

Dam planning process was once considered an area of expertise of engineers and economists. It is now an area of concern for environmentalists, anthropologists and human rights activists. They have criticized that dam projects have produced disasters. Several authors have brought into sharp focus the political, social, economic and environmental issues to which construction of dams give rise. Dam-induced human displacement has emerged a major issue worldwide. Large dams have profound effect on the landscape and on the way people live. Hoover dam supplied water and power and transformed the western US. The dam provided jobs to about 15,000 workers. However, more than 200 workers died during Hoover's construction.[19] Prior to construction of dams, rivers in deserts used to carry several tones of salt to the sea. It is now observed that this salt is spreading across the irrigated landscape, slowly poisoning the soil. The silt carried by a free flowing river enriches the downstream wetlands and deltas. The nutrients carried over by free flowing rivers help to rejuvenate fishery and marine life. After the construction of dams, this natural process is obstructed. Biodiversity loss, environmental degradation, water scarcity, social injustice, growing gap between rich and poor, difficulties faced by indigenous peoples are some of the problems associated with large dams.

The WWF report on free flowing rivers identifies the specific value of a free flowing river in contrast to one that has been obstructed with a dam or otherwise modified and urges that a concerted effort for their conservation is urgently needed. Throughout the world there are numerous examples of tribal people who live in close relationship with their rivers. The damming of the rivers affect the river's productivity. The construction of dams have resulted in the depletion of production of plants and animals or disappearance of some species, thus resulting in the loss of tribal peoples' "natural capital".[20] Researchers have reported that the completion of Farakka barrage deteriorated the flood problems of Maldah district in West Bengal. Unchecked deposition of silt on the river raised the river bed causing devastating floods in the western side of the district. The execution of Farakka barrage without proper silt management studies has resulted in choking of the lock gates due to continued deposition. This has caused breaching of riverbanks followed by flash floods. Massive bank erosion causes loss of valuable land. Roads, railways, crops, houses are destructed every year and water logging thereafter also has a negative impact.[21]

The Pak Mun dam was completed in 1995 on the Mun River in Thailand. Its Environmental Impact Assessment (EIA) study predicted a substantial increase in fish production from the reservoir. A review for World Commission on Dams (2000) suggests that Pak Mun dam had negative impacts on migratory and other fish species. The fishermen complained that their income was affected following the construction and operation of this dam. There are a number of impacts of human modification including dam building, on the life cycle of a river. Damming rivers modifies the biogeochemical cycle in different ways. Dams alter river ecosystem and have an adverse impact on fish species native to the river. Rivers contribute to pollution control through transport and removal of pollutants. This capacity is reduced because of dams as pollutants are trapped behind dams through accumulation of sediment. A conservation group of WWF has observed that the Ganga, the Indus, the Nile and the Yangtze are among the 10 most endangered rivers of the world. Indiscriminate extraction of water from their basins, coupled with the damming of their tributaries for irrigation has seriously reduced their regeneration and flow. (*The Times of India*, Ahmedabad, 21-03-07)

Dams as Tourist Attraction

When the construction of Bhakra Nangal project, a major dam project in India, was completed, Pt. Jawaharlal Nehru, first prime minister of India proudly announced that these project sites are the new pilgrimage centres of India. Like new places of worship, they became the centres of tourists' attraction. Krishnaraj Sagar dam on the river Cauveri near Mysore has become a tourist attraction with the Vrindavan gardens developed adjacent to it. It attracts thousands of tourists every day especially in the evening due to the fountains and illumination which have enhanced its beauty. The Sardar Sarovar dam at Kevadianagar in Vadodara district of Gujarat is also being developed as a tourist spot by Gujarat government, and it had been active in showcasing it as Gujarat's pride and lifeline since its construction started. The Three Gorges Dam on the Yangtze River is becoming China's largest industrial tourist destination with over one million visitors in 2005.[22]

World Commission on Dams

The World Bank is the largest financier of large dams. It has now established policies to protect indigenous peoples and formulated stricter rules for resettlement of oustees. This was possible only after the release of the comprehensive report by the World Commission on Dams. Created in 1997, it is an independent body of 12 Commissioners, charged with assessing dams' impacts – positive and negative – and providing guidelines for future construction. It has developed international guidelines for building and operation of dams to balance the competing demands of the economy and the surrounding environment. After two and half years of its formation, the Commission brought out its final report in November 2000. It is accompanied by the keynote address by Dr. Nelson Mandela. Titled "Dams and development: a new framework for decision-making", it is more than 400 pages long. The Commission report has become a standard compilation of best practices. Its first part is based on the findings of in-depth studies of impacts of existing large dams ever conducted. The report points out that "dams showed a marked tendency towards schedule delays and significant cost overruns; irrigation dams did not recover their costs, did not produce the expected volume of water and were less profitable than forecasted. Their environmental impacts are more negative than positive, and in many cases have led to irreversible loss of species and ecosystems. Millions of people have suffered by displacement. True profitability of large dam scheme remains elusive. Tropical dams emit greenhouse gases released by vegetation rotting in reservoirs."[23] According to a study published by World Commission on Dams, "all of India's dams have been planned and built without any consideration of less costly and damaging alternatives, and most have paid only cursory attention to the social and environmental disarrays they cause". The study further points out that "dams have not only helped to maintain the current inequalities in the Indian society but, in some ways, have exacerbated them" (Leslie, 2005). The second part of the report provides a framework for building dams in the future. It calls for examining cheaper and less damaging alternatives before deciding on dams. The report suggests to identify problems before dams are built, instead of afterwards.

(Prabha Shastri Ranade, Consulting Editor, Icfai Business School Research Centre, Ahmedabad.)

Endnotes

1 Leslie, 2005.

2 Down to earth, September 30, 2006.

3 Kalwar, 2005.

4 Down to earth, September 30, 2006.

5 Down to earth, October 31, 2006.

6 *Bbcnews http://news.bbc.co.uk/go/pr/fr/-2.*

7 *Bbcnews http://newsvote.bbc.co.uk.*

8 Satyanarayana, 2005.

9 Down to earth, September 30, 2006.

10 Sharma, 2004.

11 Awasthi, 2006.

12 *http://in.news.yahoo.com/070305/43/6oct02.html*

13 Sharma, 2004.

14 Satyanarayana, 2005.

15 Leslie, 2005.

16 *http://assets.panda.org/downloads/freeflowingriversreport.pdf.*

17 Wikipedia, the free encyclopedia.

18 Leslie, 2005.

19 Leslie, 2005.

20 *http://assets.panda.org/downloads/freeflowingriversreport.pdf.*

21 Betal, 2002.

22 *http://assets.panda.org/downloads/freeflowingriversreport.pdf.*

23 Leslie, 2005.

References

1. Awasthi, Kirtiman, Making adversity work, Down to earth, October 31, 2006.
2. Betal, Himanshu, "Flood problems of Maldah: A geographical analysis", *Geographical Review of India*, Vol. 64, No 4, Dec 2002, pp 337-345.
3. Forces of Nature: Floods, Down to Earth, 30 september 2006.
4. Flood hit Barmer still suffering, *http://in.news.yahoo.com/070305/43/6oct02.html*
5. Free flowing rivers: Economic luxury or ecological necessity? *http://assets.panda.org/downloads/freeflowingriversreport.pdf.*
6. Kalwar, Sharma and Kalwar, "Drought and floods in Rajasthan", *Annals of the National Association of Geographers, India,* Vol 25, June 2005, pp 81-93.
7. Leslie, Jacques, Deep Water, The epic struggle over dams, displaced people and the environment, Farrar, Straus and Giroux, New York, 2005.
8. Satyanarayana, B Geohazards, Indian Scenario, Presidential address, Section of earth System Sciences, 92nd Indian Science Congress, Ahmedabad, 2005.
9. Sharma, Archana, "Floods: A threat to sustainable development", *Indian Journal of Regional Science,* Vol. 36, No. 1, 2004.
10. State of India's Environment: A Citizen's Report – Floods, flood plains and environmental myths, Center for Science and Environment, New Delhi, 1991.
11. Strahler, Physical Geography, John Wiley & Sons, New York, 1975
12. *http://newsvote.bbc.co.uk*
13. *http://news.bbc.co.uk/go/pr/fr/-2*

2

Who takes Decision for LargeDams?
Many Questions, Few Answers

Himanshu Thakkar

This article shows in a nutshell the real picture of political economy of dam construction in India. Former Prime Ministers, Jawaharlal Nehru and Rajiv Gandhi said India is facing the 'disease of gigantism' and dams are built just to show the capability to build huge dams. Also AB Vajpayee was supporting the construction of large dams. In fact, he inaugurated many dams in his tenure as PM. TheWorld Bank is also pushing for construction of large dams. Along with dams, the river interlinking project is also being initiated. Even today, for many rivers, feasibility reports are not ready. The article says public should be involved in the decision-making process of these dams.

"At the dawn of Independence India relied, wistfully, on her high dam-builders... During this TVA phase of India's economic development, a well-known Indian engineer used to proclaim off and on that he was going to build the highest dam in

---

Source: www.infochangeindia.org The article was originally titled as "Who takes Decision for Large Dams? How? Why? Who Profits? Who Pays? Many Questions, Few Answers."

the world, suggesting implicitly a new yardstick for measuring national greatness – the height of a dam and the millions of cubic yards of concrete poured."

– Sudhir Sen.

(The first Chief Executive Officer of the Damodar Valley Corporation, called India's TVA by many) A Richer Harvest: New Horizons for Developing Countries, Tata McGraw-Hill Publishing Co., New Delhi, 1974.

The above quote is remarkable for a number of reasons. It should be noted that Sen belonged to the mainstream official group of people that were at the helm during the first decade after Independence in 1947. So this is not coming from somebody outside the system. Secondly, the "well-known Indian engineer" that Sen described here is Ayodhya Nath Khosla, who can be safely described as the first Engineer-in-Chief of Independent India (this title is used just to describe his position, it is true that no title like that existed). And lastly, the dam that is referred in the quote is the most famous icon of India's dam-building history, the Bhakra Dam.

The quote above is also remarkable because it in a nutshell reflects a reality about political economy of dam-building in India. The people who took decisions about large dams in the initial years after Independence got away with a whole series of decisions about building large dams, for which they did not have to answer any questions. Sen goes on to say about Khosla and company, "That many engineers in India, if left to themselves, like to build monuments to themselves – regardless of the time and cost involved, is a commonplace of history. India had yet to discover this." These are strong words coming from someone who occupied a very senior position in the scheme of things then.

What this means is that many of the decisions about the large dams were not taken on merit and that the people who took the decisions were not answerable in any real sense of the term. Well-known scientist of W Bengal, Meghnad Saha said in Parliament in 1954 about AN Khosla, who was the first chairman of the Central Water, Irrigation and Navigation Commission, "The chairman of the CWINC was combining in himself the functions of Brahma, Vishnu and Maheshwar. He drew up the designs, he executed the schemes himself and as Secretary he passed the whole thing himself."

But the responsibility for the decisions about these projects does not rest with Khosla alone. The politicians, the bureaucrats and the various other institutions were equally responsible for the decisions that lead to water resources development being centered around large dams. The then Prime Minister Jawaharlal Nehru himself said on November 17, 1958, while addressing the 29th annual meeting of the Central Board of Irrigation and Power, "For some time past, however, I have been beginning to think that we are suffering from what we may call, *disease of gigantism*. We want to show that we can build big dams and do big things. This is a dangerous outlook developing in India.... the idea of having big undertakings and doing big tasks for the sake of showing that we can do big things is not a good outlook at all." It is another matter that Nehru could do little to reverse the trend of unaccountable decisions.

The trend that was thus established in 1950s has continued till today as far as decisions about water resources development is concerned. Nehru's grandson and the then Prime Minister Rajiv Gandhi, famously said in 1986 while addressing a conference of Irrigation Ministers, "The situation today is that since 1951, 246 big surface irrigation projects have been initiated. Only 66 out of these have been complete; 181 are still under construction. Perhaps, we can safely say that almost no benefit has come to the people from these projects. For 16 years, we have poured out money. The people have got nothing back, no irrigation, no water, no increase in production, no help in their daily life." Rajiv Gandhi may have been off the mark about some of those numbers, but he was right on the spot about the end result. Unfortunately, these brave words did not get translated into any effective action to counter the situation.

The most glaring example in this regard in recent times is about how the river-linking scheme is being pushed. Even today, it is known that for a number of proposed links even feasibility report is not ready. There has been no proper consultation with the people, even basic information is yet to be made available to people. The assessment of costs, benefits, impacts, options, nothing is known in detail. And yet, the President of India has been repeatedly advocating these proposals for over three years now. This advocacy started in his speech to the nation on August 14, 2002 and the most recent instance when he repeated this was in his address to the nation on August 14, 2005. The Supreme Court of

India has also been pushing implementation of this scheme, though this is clearly beyond the mandate of the Supreme Court. Ramaswamy Iyer, former Secretary, Union Ministry of Water Resources has said, (*The Hindu*, December 14, 2004), "It is necessary to remind the readers that in a sense the President of India and the Supreme Court are the originators of the ILR project. Meanwhile, the President continues to commend the project in his speeches on various occasions, and the Government of India is under an obligation to report periodically to the Supreme Court on the status of the project. Those two factors indirectly act as a kind of pressure on the Government..."

When NDA government was in power before May 2004, the Prime Minister AB Vajpayee, the deputy Prime Minister, the Vice President and Ministers concerned were pushing the project that has yet to pass the first stage of due process of decision-making. With advocacy from all these high functionaries, will the due process have chance of considering the project on merit? The answer is clearly a big NO.

In fact when Vajpayee was Prime Minister, he seemed particularly favourable to large dams, as is clear from some of his actions listed below.

- **December 13, 1999:** The PM laid the foundation stone for the 800 MW Parbati Hydel Power Project (stage II).
- **June 5, 2000** (World Environment Day): The PM lays Foundation stone of 800 MW Kol Hydel Power Dam in Himachal Pradesh.
- **March 4, 2001:** He dedicated the Thein dam (Punjab) to the nation.
- Launching of 50,000 MW hydropower initiative.
- Launching of Task Force on river-linking programme.
- Ordering closure of Tunnel 3 & 4 of the Tehri Project, even as the people to be affected were yet to be resettled.
- Supporting and pushing for increase in height of Sardar Sarovar Dam even as the people to be affected were yet to be resettled as per legal requirements.

- Pushing for the Indira Sagar Project, leading to submergence of over lakh people including those in Harsud Town and over a hundred villages, without due rehabilitation.
- Formulating National Resettlement and Rehabilitation policy without legal teeth and with just provisions, much diluted when compared with the R&R norms of say Sardar Sarovar Project.

All these (and some other) actions collectively gave a big push to taking up big dams at an accelerated pace. When each of those projects involves hundreds of crores of rupees, and in some cases thousands of crores, one can understand the kind of benefits and perks that the various arms of the government would reap.

Another actor that has been playing an important role in pushing projects without justification is the World Bank. The WB may have directly funded only a small fraction of the large dams taken up in India, but it has played a significant role in shaping India's water resources policies and programmes and pushing such projects. This it has done many times in violation of its own policies and in violation of Indian policies. For example, when the World Bank signed the agreements to fund the Sardar Sarovar Projects in 1985, the Ministry of Environment and Forests (MEF) had yet to give clearance to the project from environmental point of view and for diversion of forest land. In fact the agreement signed with the Bank and the dues that Indian government had to start paying for the same were used to put pressure on the Ministry of Environment and Forests to give clearance to the project even when the Ministry explicitly said that the project is not ready for clearance. In fact, the World Bank agreement was used as a certificate of merit to push various clearances for the project.

The World Bank similarly pushed the Nathpa Jhakri Hydropower project in India without full consideration of all the costs, benefits and impacts of the project and signed an agreement to fund the project in 1988-89. The consequences of that are visible today, when the project had a time overrun of over 7 years and cost overrun of over 358% over the projected cost. Moreover, the project is facing serious problems in terms of flash floods, the siltation leading to stoppage of power generation, and so on. So much so the Parliamentary standing committee on energy issues, in a report tabled in Lok Sabha on August 18, 2005, has recommended a thorough enquiry into these issues.

Allain Duhangan Hydropower project in Himachal Pradesh is another instance where the World Bank (this time International Finance Commission, the private sector arm of the World Bank) was found guilty of pushing the project without adhering to its own norms and in fact based on falsehoods. In October 2003, IFC declared on its website that the project would go to the WB Board on October 31, 2003, and that all the necessary requirements including the Environmental Impact Assessment and the Public Consultation has been completed. When SANDRP (South Asia Network on Dams, Rivers & People) along with the people from affected villages exposed that these were lies, the project had to be withdrawn from being taken to the Board. But the Bank seems to have learnt no lessons and it pushed the project through the Board a year latter after paying some lip-service to the concerns of the people, without even completing the EIA in all respects.

One fundamental problem is that there is no participatory bottom-up decision-making process that would help put in a sense of transparency and involvement of the local people in the decisions. Moreover, even after completing about 4,000 large projects, there is no credible comprehensive assessment of performance of any large dam project. These factors have helped the system to go on even in face of big gaps between promises and performance.

What is shocking is that the authorities have found it difficult to adhere to even whatever little processes that are required today for people to participate in the decisions about projects. For example, the project authorities have been using various measures to scuttle the effective participation of the people in the public hearing process that has become mandatory since 1997 before any large project can be cleared. The quality of most of the EIAs that have become mandatory for large projects since 1994 is shameful. The Ministry of Environment and Forests, that is supposed to be responsible for these processes, has been acting not only as a rubber stamp, but many a time has been actively involved in helping the project authorities scuttle the due process.

The latest and most shocking instance in this regard came to light in March-April 2005, when the MEF gave clearance to the proposed Chamera III and Parbati III hydropower projects in Himachal Pradesh. The EIA notification requires that the report of the public hearing for such projects should come to the MEF with inputs from the State Environment Department and State Pollution Control Board. However, before the state government could take a decision about these projects, the MEF gave clearance to the projects, apparently under pressure from National Hydroelectric Power Corporation. The additional Secretary, Department of Environment, Himachal Pradesh strongly protested in his letter dated April 25, 2005 to the Secretary, MEF, "These decisions have serious ramifications as it is State Government and the inhabitants of the area who have to directly bear the brunt of environment-related problems. Moreover, the role of the State Government and the State Pollution Control Board as enunciated in the EIA Notification (and related clarifications) as enshrined in Article 48A of the Constitution of India has been apparently overlooked."

When the report of the World Commission on Dams was made public in November 2000, a golden opportunity was available to make some significant changes in the decision-making process of water and power development projects. However, senior officers of the World Bank worked over time to ensure that a number of key developing country governments opposed the WCD recommendations and then the Bank used such opposition to justify non-adoption of the recommendations in the Bank policies.

The only ray of light in these dark surroundings is that at many projects people are becoming conscious of the serious adverse impacts of large dams and are evincing opposition to unjustifiable projects. It would, of course, take a long time before these responses are strong enough to counter the strong entrenched interests pushing large dams.

(Himanshu Thakkar (ht.sandrp@gmail.com) is Coordinator of the South Asia Network on Dams, Rivers & People (SANDRP) and Editor of "Dams, Rivers & People".)

3

Interlinking of Rivers in India – Assessing the Justifications

Jayanta Bandyopadhyay and Shama Perveen

The present form of the river interlinking proposal, made by the National Water Development Agency, has been hailed as a 'must' for the country by many politicians. This paper critically examines the assumptions behind and the main justifications extended for the project. The paper disagrees with the concept that river basins can be mechanically divided as 'surplus' or 'deficit' ones, and views the proposed interlinking as an extremely cost-ineffective measure for the expansion of a rather inefficient traditional irrigation process. Thus, in the event of the mega-project being taken up as it is, it will lead to sub-optimal use of the water resources of the country through a huge and unwise investment. The official justifications for the proposed interlinking of rivers are not found to be backed by any scientific reasoning.

I. Water Resources of India

In spite of the surface of the earth being covered mostly by water, over the past few decades water scarcity has emerged as a global problem. All life forms and

Source: Economic and Political Weekly, December 11, 2004. (www.epw.org.in/). © EPW. Reprinted with permission.

human economic activities are critically dependent on water, the movement of which is governed by the global hydrological cycle. Humans have moderated such movements of water and made available large quantities of this resource at times and places to suit the needs of societies and meet the demand of economies. In course of time, the level of such human interventions has grown to such an extent quantitatively that the need for a more informed approach to this vital natural endowment is being recognised and articulated all over the world (Falkenmark *et al.* 1980; Wittfogel, 1957).

At the global scale, "freshwater lakes and rivers, which are the main sources for water consumed by the human societies, contain on an average about 90,000 Billion Cubic Metres (BCM) of water, which is about 0.26 percent of total global freshwater reserves" (Shiklomanov, 1993). The global withdrawal of freshwater has increased sevenfold between 1900 and 2000 (Gleick, 2000). If this trend continues, the world may see a more than sixfold increase in the number of people living in conditions of water stress, from 470 million today to 3 billion in 2025 (Postel, 1999). With respect to its share of global water resources, India is regarded as a better endowed country, with a total annual precipitation of about 4,000 bcm. This is about 4 percent of the total average annual runoff in the world's rivers (NCIWRDP, 1999a). As reported by the NCIWRDP (1999a), India has annual available water resources of 1,953 cu km. On the other hand, it is also true that, if the population of a country grows rapidly, there will naturally be a proportional reduction in the per capita availability of water. From this point of view, India is facing a regime of stress (Dyson, 2001), as the per capita availability of water declined from around 5,177 cu m in 1951 to 1,869 cu m in 2001. Given the projected rise in India's population by 2025, the per capita availability may drop further to below 1,000 cum. In terms of country-level indicators, such a situation would be considered as one of 'water scarcity'. This picture is not typical of India, but a rapid growth in population exacerbates the situation.

Water Resource Scenario

In terms of the total annual precipitation per unit of land surface, India stands well above the global average. However, the figures of gross annual precipitation over the country as a whole provide an unrealistic picture of the actual situation with respect to availability of water in the country. The domination of the southwest monsoon in the making of the climate of south Asia results in a wide spatial variation in the levels of precipitation from the east to the west (see the figure) and an acute temporal variation through the concentration of heavy precipitation during the 2½ months of monsoon, spread over July, August and September. The variability is exemplified by a comparison of the number of rainy days in parts of Rajasthan along the north-western boundary, where it is just five, and in some areas in north-east, where it rains for about 150 days (NCIWRDP 1999a:12). Similarly, the average annual precipitation ranges from a low of about 200 mm in some locations in the western regions of India to a high of about 11,000 mm in the north-east.

It is thus quite clear, that the averaged-out national picture of precipitation cannot provide any realistic picture of the actual problems of water resource development and management. Due to the spatial and temporal inequities in precipitation, rivers in most parts of the country show a skewed hydrograph. The

Spatial Variation in Precipitation Pattern

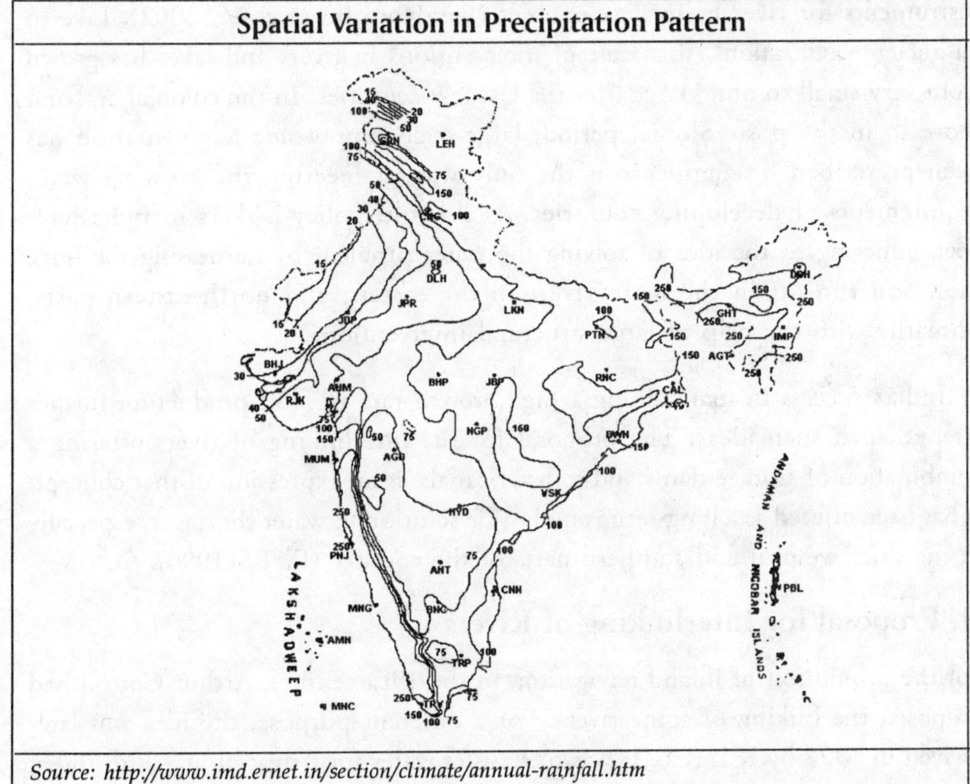

Source: http://www.imd.ernet.in/section/climate/annual-rainfall.htm

rain-rich regions experience regular annual inundations during the monsoon, while many rivers in the lower rainfall areas dry up soon after the monsoon is over. There is an increasing tendency to describe such obvious hydrological unevenness of rivers as disasters. Whenever a river overflows its banks, statements accusing 'too much rainfall' or 'release of excessive water from the upstream' have become common. They are, however, not backed up by any open data. In areas with lower precipitation, growing water scarcity resulting from unsustainable utilisation practices and vanishing conservation measures, are frequently blamed on 'droughts'. How much of the growing scarcity is the result of human interventions in the water systems and how much is due to meteorological drought has not been openly analysed (Bandyopadhyay 1989).

As humans increasingly started to intervene in the flow of rivers (Deursen 2000), the gradual success of these interventions resulted in the upscaling of their operational dimensions that finally led to the emergence of large dams as instruments for river basin development (Bandyopadhyay *et al.,* 2002). Like in all ancient civilisations, this scale of interventions in rivers and lakes has grown from very small to quite large over the last few centuries. In the colonial era, and more so in the post-colonial period, large-scale supply-side augmentation has been prescribed by engineers as the only way of meeting the growing water requirements of developing countries. As a result, policy-makers in India have been adhering to the idea of solving the water problem by harnessing the huge monsoon run-off in the main rivers in the eastern and north-eastern parts, primarily with the help of large structural interventions.

India's success in maintaining a high growth rate in food production further strengthened such ideas. The proposal for the interlinking of rivers offering a combination of storage dams and transfer canals, is an expression of that concept. It has been offered as a long-term supply-side solution to water shortages, especially in the drier western and southern parts of the country (IWRS 1996).

II. Proposal for Interlinking of Rivers

For the promotion of inland navigation, in the 19th century, Arthur Cotton had proposed the linking of some rivers. For a different purpose, the idea was later revived in 1972 by K L Rao. It was to transfer water for irrigation in south India, where demand was quickly outstripping availability. The result was the idea of the 'Ganga-Cauvery Link Canal' as proposed by Rao (NCIWRDP 1999a:179-80). In 1977, Captain Dastur, an aircraft pilot, proposed 'an impressionistic' plan for the construction of a pair of canals. Better known as 'the garland canal' scheme, it envisaged the construction of a 4,200 km long Himalayan canal and 9,300 km long southern garland canal and the connection between the two systems through two pipelines passing by Delhi and Patna.

On the basis of studies undertaken by the Central Water Commission (CWC) and other expert bodies, these two proposals were not found to be worthy of serious consideration. However, in August 1980, the ministry of water resources framed a national perspective for water development, and the National Water Development Agency (NWDA) was established in 1982, to carry out studies in the context of the national perspective. The perspective has two main components: Himalayan river development and peninsular river development.

Under this perspective, the NWDA took up the task of developing a proposal for interbasin transfer of water that would be more comprehensive than previous ones. The proposals of NWDA for long-distance interbasin transfers have not been openly articulated so far with any technical details. In fact, it is reportedly still at the stage of an idea, and not a project. The NWDA was to survey and investigate 'possible storage sites and interconnecting links' in order to establish the feasibility of proposals forming part of the national perspective. According to the report of the NCIWRDP, the interlinking proposal aims to provide large-scale human-induced connectivity for water flows in almost all parts of India, through a total of 31 links on both the Himalayan and peninsular components. None of the pre-feasibility or feasibility studies are available in the public domain for an independent professional assessment of the technical and economic justifications for these links. However, the project has been described by many politicians as a win-win situation for the mitigation of floods and droughts. The issue gained sudden and renewed currency in political, legislative and civil domains when the Supreme Court, in connection with a public interest litigation, passed an order on October 31, 2002 for the government to complete all links of rivers within 12 years.

Justifications Advanced

The idea of interlinking rivers is described, from the president down to local-level leaders in regions with lower rainfall, as the perfect win-win solution that will address the twin problems of water scarcity in the western and southern parts of the country and the problem of floods in the eastern and north-eastern regions. The claims and statements of politicians, cannot, however, substitute comprehensive scientific and technical assessment so that one can know whether, if the proposed interlinking project is completed, the right quality and quantity of water would be stored and delivered at the right time in the rightplaces. Further, all this is to be achieved in the most cost-effective manner, without creating any social deprivations like inadequate rehabilitation of the involuntarily displaced. In view of the fact that 'our past record in resettlement is deplorable', as noted by the former chairman of the task force for interlinking of rivers (Prabhu, 2004), nothing less than an open professional assessment of the technical proposals based on the latest interdisciplinary systems knowledge would be appropriate.

Unfortunately, beyond a few lines drawn on the map, no scientific and technical information of the proposal is available in the public domain, so that an open and independent professional assessment can be undertaken. As a result, the justifications that are being presented to promote the project have remained mere exercises in guessing by the professional world as well as the common people. This absence of open professional assessment of the technical details of the proposals has important implications. For example, in contrast with the official prescription for interlinking of rivers, there is a strong opinion that India's water crisis is caused by the mismanagement of water resources, and the solutions do not lie merely in supply-side augmentations. Iyer (2000), a former secretary of water resources, corroborates this fact, with the view that:

> *There has been no serious attempt to work out a series of area-specific answers by way of local conservation and augmentation to the maximum extent possible. The severe drought of the summer of year 2000 in India was not an indication of water 'insecurity' nor did it point to the need for big projects or long-distance water transfers... The drought conditions were a result of bad water management in the past and the answer lies in better resource management in the future.*

Even if the publicised claims of the gains to be made from the interlinking project are not available in the public domain, it will be necessary to examine the scientific credibility of certain concepts that are leading to these justifications. It is important to ensure that such a costly project is not based on an outdated and questionable knowledge base. In particular, there is a great need to take into full cognisance, the move towards a new and interdisciplinary paradigm for looking at water resources the world over. Thus, the proposed interlinking of rivers needs to be assessed in the light of the emerging paradigm changes in the subject of water management.

Changing Paradigm of Water Management

The urgent need for a change in our vision of water has, over the past few years, found frequent expressions in the scientific literature on water and its management. The caution that the 'business as usual' way of looking at this crucial natural resource would lead to severe stress, and probably conflicts, has been registered even at the highest international professional platforms (WMO 1992:27; Cosgrove and Rijsberman 2000:xxi). Similar concerns have been expressed in diverse contexts by many leading water professionals over the past several years (Biswas 1976; Falkenmark *et al.,* 2000; Gleick, 1998; Wolff and Gleick, 2002:1-32). While, on the basis of the ground realities, the need for a change in our perceptions about water has been accepted by many, the nature and extent of the required changes are still to be clearly articulated. Nevertheless, it is also apparent that what is needed is a fundamental shift away from the present reductionist engineering paradigm to a holistic and trans-disciplinary one. The new ways of managing water on the basis of such a knowledge base has increasingly been identified as Integrated Water Resource Management (IWRM).

Twentieth century water resources planning generally relied on linear projections of future populations, per capita demand, agricultural production and levels of economic productivity (Gleick, 2000). The vision of the water resource planners was limited mainly within supply-side solutions. However, the professional views of water are changing rapidly, based on the scientific analyses of past mistakes and availability of new information. This 'changing water paradigm' (Gleick, 1998) represents a real shift in the way people think about water use. For example, in the US, the country that started the old global trend of building large dams, today "there is a new trend to take out or decommission dams that either no longer serve a useful purpose or have caused such egregious ecological impacts so as to warrant removal. Nearly 500 dams in the US and elsewhere have already been removed and the movement towards river restoration is accelerating" (Gleick, 2000). Continued investments in huge engineering interventions is being challenged by those who believe that a higher priority should be assigned to projects that meet basic and unmet human needs for water (Gleick, 1996). Following these paradigmatic shifts in notions worldwide, various other means to conserve water instream are becoming evident. The Murray-Darling Basin Commission in Australia is seriously contemplating extending financial encouragement to farmers for saving on their allocation of irrigation water and to allow the savings to remain instream. In another instance, Chile's national water code of 1981 established a system of water rights that are transferable and independent of land use and ownership. The most frequent transaction in Chile's water markets is the 'renting' of water between neighbouring farmers with different water requirements (Gazmuri, 1992). In the Indian context, Bandyopadhyay (2004), in his assessment of the new National Water Policy 2002 finds that the old paradigm is still strongly entrenched in the official water administration. There are, however, some others who believe that there is no problem with the present paradigm, which is able to address all problems. However, pointing out the exclusively supply-side solutions offered by the present paradigm, Helming and Kuylenstierna (2001) have cautioned that:

> *Tapping into new supply sources tends to either impinge on demands by others, or cause serious damage to nature. Each new source of water is also normally more expensive to develop than the previous one. Demand-side management is therefore slowly becoming a new paradigm for water governance.*

Reductionist Roots of Concept of 'Surplus' Basins

The proposal for interlinking of rivers is seen in this paper as a simplistic supply-side solution put forward from the existing paradigm of water management, which is fighting for its life elsewhere in the world. The proposed project is critically dependent on the identification of some river basins or sub-basins as 'surplus' ones, from which water may be transferred to some others, identified as 'deficit' river basins. Thus, it is claimed, swapping of floods in the surplus basins with the scarcity of water in the deficit basins would be a win-win solution. The interlinking of rivers will put to use 'the water otherwise going waste in the surplus river basins' (NCIWRDP 1999a:181). Prabhu (2003) makes a more direct statement in this regard when he claims, "the (interlinking) project is about rationalisation of water that is lost to the sea".

The reductionist engineering concepts of water have seen it mainly in the form of visible flowing water. The totality of the ecosystem services provided by water, from the time of a drop falling on the surface of a river basin to the moment of its flowing to the sea have remained marginal and neglected for a long time (Falkenmark and Folke, 2002). As a result, it is not possible for the existing paradigm to recognise and record these various ecological processes and their values, for instance, in the conservation of biodiversity, its role as a mobile solvent, the pushing of the sediment load out to the sea, and many others. It is this conceptual limitation of the present paradigm that makes it possible for it to describe the outflow of a river to the sea as a 'waste', or finds little difficulty in locating 'surplus' river basins in a limited arithmetic assessment. Thus, the NCIWRDP (1999:181) declares that:

> *...creation of storages and interbasin transfers from the surmised surplus river basins to deficit basins has been the guiding objective... it is considered imperative that all the rivers in the country be linked by a national grid to meet the shortages in the various parts of the country.*

The methodology for working out whether or not a river basin has any 'surplus' water, is based on an unpublished paper by Mohile (1998) and recommended in the report of the working group on interbasin transfer of water (NCIWRDP 1999b). In following this approach, a simple exercise in arithmetic hydrology has been employed that externalises the whole set of ecosystem services provided by water in a river basin. Whilst the process for the assessment of the needs of irrigation, domestic and industrial sectors have been dealt within the report (NCIWRDP 1999b:30), there is no information given on how the water needs for the continuation of diverse ecosystem services provided by water in the various parts of the basin would be assessed. Leaving such parameters out may be convenient for promoting a project, but, as exemplified by Costanza *et al.* (1997), the arithmetic may be in conflict with the total reality. The exercise by Costanza estimated that the value of the world's ecosystem services could be almost twice as large as the global gross national product (33 trillion per year compared with around $18 trillion per year). In this regard, Vaidyanathan (2003) rightly argues that simply identifying a river basin on the volume of flood flows is a misleading basis for judging surpluses. Bandyopadhyay and Perveen (2003) have described such a reductionist and mechanical approach to the categorisation of river basins as 'arithmetical hydrology'.

Thus, when the reductionist vision of arithmetical hydrology is replaced by the holistic perspective of ecohydrology, the outflow of a river to the sea is no more seen as a 'loss' but as essential for the continuation of the ecosystem services of the river. Whilst floodwater in eastern India is seen as a 'harmful surplus' from the viewpoint of arithmetical hydrology, the same floodwater is seen as a source of free minerals for the enrichment of land, free recharge for groundwater resources, a free medium for the transportation of fish and conservation of biological diversity and free bumper harvest for humans, from the ecohydrological viewpoint.

Cautionary statements that diversion of water from the 'surplus' to the 'deficit' basins would have significant impacts on the physical and chemical compositions of the sediment load, river morphology, aquatic biodiversity and the configuration of the delta, are not new. However, their assessment is not possible from within 'arithmetical hydrology'. These downstream processes have serious economic and livelihood implications and it is a national imperative to address and assess them as integral to the benefit-cost analysis of the proposal. Considering the diverse factors involved, Singh (2003), who chaired the working group on interbasin transfer of water of the Ministry of Water Resources (MOWR), takes the position that:

There really seems to be no convincing argument or vital national interest, which can justify this mammoth undertaking (interlinking), in its entirety.

All river basins have evolved over the geological past by making optimal use of available water resources. There are no inherently 'surplus' river basins. It is convenient for the reductionist viewpoint to ignore the totality and the fact that all drops of water in all river basins at all times are performing several ecosystem services. As a result, any transfer of water from one basin to another is not a simple arithmetic exercise, but one of a complex impact assessment.

However, there may always be a good case for interbasin transfer. This cannot be done in an ad hoc manner, without recognising the diverse social, economic and ecological impacts of such transfers. Further, full economic compensations for those affected by the negative impacts should be fully made. In the available information on the proposal for interlinking of Indian rivers, there is no reference to the assessment of any such costs. Accordingly, it is apparent that there are many important reasons for having a close look at the scientific validity of the justification put forward for the proposed interlinking of rivers. For this, all the pre-feasibility, feasibility and detailed project reports regarding individual interlinking proposals need to be made available in the public domain for open professional examination.

III. Assessment of Justifications

As has been mentioned above, beyond some lines drawn on the map, no scientific and technical details of the proposed interlinking of rivers is available in the public domain. Technical information on the flows, storages, link canals, barrages and associated engineering structures, the ecological impacts on downstream areas of the basins, extent of involuntary displacement, and likely costs and benefits of the proposal have not been made available for open professional assessment by the ministry of water resources. However, judging from the media coverage of public presentations of politicians, there seems to be a belief that after this project is completed, there will be no shortage of water in the rain-scarce parts of India, and that the 'problem' of floods in the rain-rich parts will be solved. Information on the mega projects on water with such great positive impacts have, however, remained guarded against any transparent and independent professional assessment (Bandyopadhyay 2003). The importance of the people is limited to paying the huge costs of the project, whose justifications they have to just assume and to accept a standard of resettlement that is 'deplorable' (Prabhu 2004).

Large water projects have usually been justified by considering the direct cost of construction and operation of the engineering structures. Such practices have invariably generated popular discontent and have disgraced water-related project in the public eye. Careful analysis and evaluation of all costs (intrinsic social, economic and environmental), therefore, becomes a *sine qua non* for a more informed decision-making. On the basis of the very general description of the proposal as available in the public domain, such a detailed assessment cannot be taken up. What can be undertaken is merely an assessment of the justifications given for the investment and how the proposal for interlinking stands against the other available option for providing the claimed benefits. This section examines some of the important justifications, believed to be the pillars on which the very costly project of river interlinking is being placed. The identified questions are:

- Does the proposed interlinking of rivers offer the most cost-effective solution to the scarcity of water in the drier parts of the country?
- Does the food security of the common people of India depend on the proposed interlinking of rivers?
- Who would bridge the knowledge gap in the Himalayan component?

Issues of Cost-Effectiveness

Satisfaction of domestic water needs should receive top priority in policy and be seen as a part of human rights (Gleick, 1999; McCaffrey, 1992). Domestic water requirements are of the highest priority in terms of supply. However, in terms of quantity, these requirements are small compared with that for irrigation. Surprisingly, the official method for the identification of 'surplus' or 'deficit' river basins, clubs together all the various water requirements, thus obscuring the priority attached to separate water needs and demands. If the objective of providing domestic water security is given the highest priority, and is not clubbed with irrigation or industrial requirements, most areas in India would probably come out as self-sufficient. Nigam *et al.* (1997) had undertaken systematic case studies in a few areas of the country considered to be water scarce. Their study makes it clear that if the precipitation available within the respective watersheds or sub-basins is harvested and conserved properly, the satisfaction of domestic water needs would not be a problem in most parts of the country. For promoting domestic water security, local level water conservation is a suitable option compared with developing large storage and long-distance diversion facilities, as these often carry high financial, social and ecological costs (World Commission on Dams, 2000). This observation is completely in consonance with the results of numerous community initiatives for water harvesting in India, whether in Maharashtra, Gujarat, Rajasthan, Tamil Nadu, Uttaranchal or anywhere else in the country.

The problem of scarcity gets further compounded by the growing tendency in India to hurriedly get short periods of scarcity or inundations identified as 'droughts' and 'floods', clearing the way for obtaining funds for relief. There is little concern about whether these 'natural disasters' are the results of natural extremes or of human interventions. Bandyopadhyay (1989) has analysed the risks associated with such opportunistic identification of human induced water scarcities with naturally occurring drought conditions. Under climatic conditions dominated by the monsoon, droughts and floods are quite expected events. Moreover, though fluctuations in the precipitation cause one form of drought, there are also many other forms of drought that lead to water scarcity. These distinctions of diverse forms of drought are reconfirmed by R R Kelkar (*Businessworld,* 2001), when he warns that:

> *If the rainfall over a given region is more than 25 percent below normal, meteorologists call it a drought. However, this does not always bring out the true picture since crops could still survive if they get enough rain at the critical growth stages. On the other hand, a statistically normal rainfall but with a few spells of very heavy rain interspersed with long dry spells can cause agricultural drought as opposed to a meteorological drought.*

The impression that this mega-project offers all the solutions to the problems of water scarcity in India has diverted public interest from promoting local-level initiatives for the harvesting and conservation of water. A belief has been created that there is 'enough' water in the 'surplus' rivers to cater to all the needs of all the people. What is to be done is only to invest in this mega-project for transfering water from one basin to another. This is why, when a higher than average rain falls on water scarce states like Rajasthan, as has been the case during the monsoon months of 2003 and 2004, serious official plans for rainwater harvesting are difficult to come by. Non-governmental initiatives, however, have achieved a great deal in this direction, establishing that local and cheaper options do exist for providing domestic water security in drier regions. Accordingly, Verghese (2003) has pointed out that:

> *The interlinking project is not a single stand-alone panacea for the country's water problems but the apex of a progression of integrated micro to mega measures in an overall but unarticulated national water strategy.*

There is another factor that needs to be addressed. The supply of domestic water needs in coastal India may soon become more economic through emerging technologies like desalination. The cost of water from desalination plants is going down significantly. It offers great hope for the people living near our long coastline. It may not be unrealistic to say that in the coming decades, there may not be much use of water supply through the interlinking project, as supplies from desalination plants would be forthcoming. Large coastal cities like London have already made massive plans for future water supply based on this technology.

As it stands, for domestic water security in either the uplands or the coastal belts, cheaper and local opportunities are available. Large inland urban areas like New Delhi and industrial towns may still need water through interbasin transfer. For addressing such situations, a clear national plan may be undertaken, with a clear pricing strategy for such supplies. This should not be mixed up with the requirements for irrigation. The case for provision of more water for irrigation, which is the central purpose of the interlinking project, should be assessed separately, especially in respect of whether the food security of the people of India is dependent on it.

Food Security

According to projections, India's population will continue to grow for some more decades and is expected to flatten towards the middle of the century. Food requirements will follow the trend in population growth. Assessment of foodgrain requirements can be made with per capita consumption and total population. India has the largest irrigation network and second largest arable area in the world. Per capita foodgrain production has been the most common indicator of food security. Recent agricultural statistics reveal that with improvements in farming technologies and plant genetics, India has achieved a record foodgrain production of 211.32 million tonnes in 2001-02, which is 15.40 million tonnes more than that of the previous year (MoA, 2003). Between 1950 and 2000, annual cereal production per capita rose from 121.5 kg to 191.0 kg (Hanchate and Dyson, 2004:229). Food security is, however, not a matter of arithmetics alone. It depends both on the quantitative availability and equitable access. It needs to be remembered that in spite of the large buffer stock of foodgrains, an

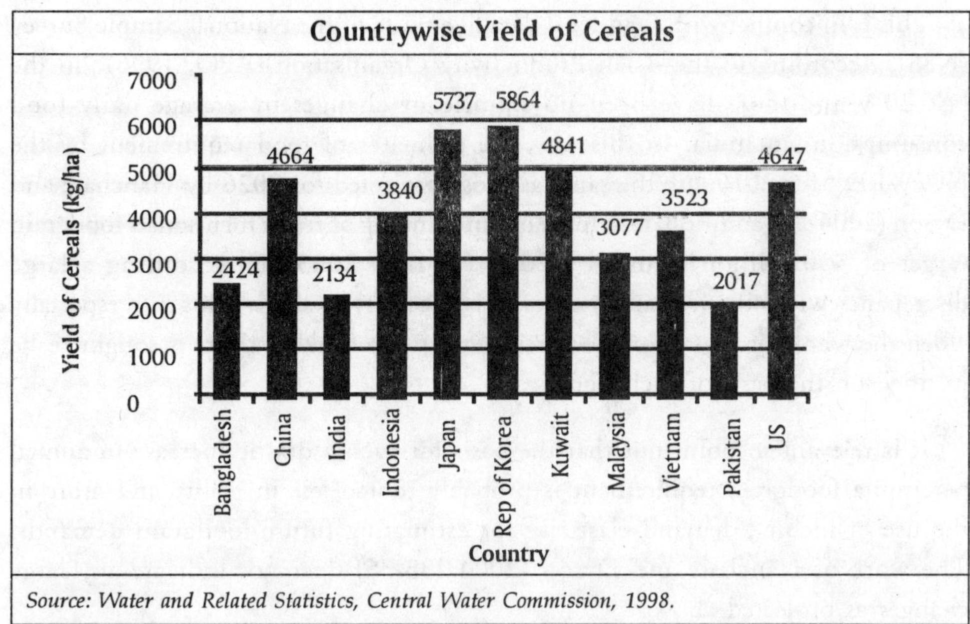

Source: Water and Related Statistics, Central Water Commission, 1998.

estimated 200 million people in India are underfed and 50 million are reportedly at starvation levels (Goyal, 2002).

Food security and foodgrain requirements: One important and probably the most powerful justification of the proposed interlinking of rivers is the perceived link of the project with India's food security. No scientific justification for this has been given separately by the task force on interlinking of rivers. Based on population projections, socioeconomic and demographic changes, and assumed changes in the pattern of food consumption, the NCIWRDP had estimated total foodgrain demand in 2010, 2026 and 2050 at high and low growth rates. These projections were made on the basis of an unpublished work by Ravi (1998), which results in a great increase in the projected foodgrain demand. While, according to the estimates of NCIWRDP, foodgrain demand in India (direct and indirect) in 2010 under the low and high demand scenario is 245 and 247million tonnes respectively, Hanchate and Dyson (2004:241), in a systematic review of past work say that:

> *...this analysis suggests that in 2026 direct cereal demand will be roughly 220 mmt, with another 30 mmt being needed for other uses, giving a 'ball-park' total of 250 mmt.*

This is in conflict with field-level data like those of the National Sample Survey (NSS). According to the Asian Productivity Organisation (APO, 1996) "in the last 20 years, there have been no significant changes in average daily food consumption" in India. In this way, the estimates of food requirement by the NCIWRDP for 2010, are the same as those projected for 2026 by Hanchate and Dyson (2004). In addition, task force on interlinking of rivers mentions a foodgrain target of 550 million tonnes by 2050 (TFILR, 2003:11), recording a large discrepancy with other research results. It is necessary to analyse this gap, especially when the whole proposal for the investment in interlinking rivers is sought to be justified on the basis of such figures.

It is relevant to point out that the basis for such a drastic increase in annual per capita foodgrain requirement is probably rooted less in reality and more in the use of income demand elasticity for estimating future foodgrain demands. The work of Hanchate and Dyson (2004:233-35) does not indicate any large changes as projected above:

> *According to the NSS, annual consumption of all cereals combined fell from 175 kg per person during 1972-73 to about 147 kg during 1999-2000. The FBS (FAO) figures, however, suggest that consumption rose slightly from around 153 kg in 1972-73 to about 157 kg in 1993-94, before increasing to 164 kg in 1999-2000...The NSS data on per capita food consumption underpin the projections because they provide the only state level figures.*

While making projections of cereal demand for India in 2026, Hanchate and Dyson (2004:237) accept that:

> *Accordingly, here we have simply assumed that for the rural and urban populations of each state, levels of per capita consumption will remain constant, as in 1993-94. For all India, this corresponds to annual consumption of 154 kg per person – a figure which is almost identical to the average of the NSS and FBS estimates for 1999-2000.*

Yield improvement as a measure of food security: One important factor related to food production and security is the yield. In spite of the availability of good water and land, India's agricultural productivity is very low compared with that of other countries. Yield improvement is as much an important factor in food security as expansion of irrigation. China faces challenges similar to India in terms of food production, probably more acutely. It has a larger population to feed, with much less arable land. As Swaminathan (1999:73) has pointed out, China produces 13 percent more food grains per capita than does India. Data from the FAO (CWC 1998: 223-24) indicates that while cereal yield in India stood at 2,134 kg per ha in 1995, that of China was 4,664 kg per ha (see the chart).

Agricultural scientists in India are foreseeing great technological breakthroughs that would push agricultural productivity upwards in the coming years. The NCIWRDP (1999a:57) has pointed out that wheat yield in experimental farms in India is already over 6,000 kg per hectare. However, the calculations of India's food production in the coming decades, which have been done to show the interlinking project as an essential step for food security, are based on the assumption that even 50 years from now, India will have attained field-level yield levels that are only two-thirds of what has already been achieved on experimental farms. The NCIWRDP (1999a:57) has assumed yield levels of 4,000 kg per hectare in irrigated land as the basis for making the projection for crop production in 2050.

Similarly, in rain-fed land, NCIWRDP has projected that food crop yield is expected to grow from the present 1,000 kg per ha to 1,500 kg per ha only 50 years. However, Singh (2003) takes the view that India is "already producing enough food; production can be further increased by at least 25 percent from existing irrigated area itself by improved inputs and agricultural technology". Carruthers and Morrison (1994) reiterate this view, when they say that:

> *We do not anticipate or call for an increased rate of capital-intensive investment in irrigation infrastructure but we do need to see that more is achieved with what is presently developed.*

It is important to note that China, with only half as much arable land per capita as India, is today not thinking in terms of drastically increasing volume of water in agriculture, but increasing the water use efficiency in existing irrigated areas. Wang (2002:15,110), the water resource minister of China, writes that:

> *Irrigation is no longer 'watering the land' but supplying water for growth of crops ...At present, the average agricultural water use efficiency is 0.43 in China. If water saving irrigation is extended to raise the figure up to 0.55 (some experts consider 0.6), food security can be guaranteed when the population increases to 1.6 billion in 2030 without increase of total agricultural water use.*

In India, which is blessed with more arable land and more irrigation potential, while similar figures for improvement in efficiency of the use of irrigation water (from 0.35 at present to 0.60 in 2050) have been projected on paper (NCIWRDP 1999a:58), there is no clear policy perspective to achieve higher water use efficiency and reaching the declared targets. The lack of interest in end-use efficiency in irrigation will push the farmers to the soft but costly solution offered by the interlinking of rivers. Swaminathan (1999:93) has thus cautioned that:

> *The inefficient and negligent use of water in agriculture is one of the most serious barriers to sustainable expansion of agricultural production. Public policy regarding the cost of water supplied by major irrigation projects and low cost or free distribution of power for pumping underground water aggravate the problem... Water consumption can be reduced radically, by as much as five-to-ten fold, at the same time significantly increasing crop yields.*

Vaidyanathan (2003), who has examined the methodology and estimates in the NCIWRDP report, questions the very concept of this efficiency underlying the measures. He says that:

> *The present available efficiency of surface irrigation, according to the figures cited in the report, ranges between 30 and 50 percent... The concept of efficiency not being specified, their relation to projections cannot be verified without comparable estimates of current and future water balances and irrigation efficiencies overall for the two major sources separately.*

The World Bank irrigation sector report on India takes a similar view on irrigation and says, "from the past heavy emphasis on physical expansion, effort now needs to turn to a much greater emphasis on productivity enhancement" (World Bank 1999:11). It is clear that the further physical expansion of irrigation is neither needed nor is it the most cost-effective option for maintaining India's food security.

This makes it clear that continued food security does not depend on the proposed expansion of irrigation potential by another 35 million ha through the interlinking of rivers. There are several other, more cost-effective ways to sustain food security. Hence, it is a national imperative that the costs for maintaining food security along all possible technological options are examined in comparison with the claims of the proposed interlinking. Particular attention should be given to the removal of the obstacles to technological changes in agriculture, because even after 50 long years, future yield from India's irrigated fields has been assumed to be only two-thirds of what has already been achieved on experimental farms (6,000 kg per ha).

In addition, there is a case for the review of the use of the irrigation potential already created and projects that have remained incomplete. Till the end of the Nineth Plan, the irrigation potential created and utilisation achieved in India was reported at 106.6 million ha and 93.4 million ha respectively (NCIWRDP, 1999a:79).

The reasons for taking the figure of 77 million ha as the projected irrigated cropped area as far away as in 2010 needs to be examined from the point whether it is a conservative figure. It is quite logical for the country to ensure that the irrigation potential already created be realised with the projected high level of efficiency. In such a situation, there will not be any need for the proposed interlinking of rivers, as far as basic food security is concerned.

Of course, there is another side to all this. The ready availability of additional water transferred through the interlinking of rivers would promote water-intensive commercial crops in dry areas. Such crops could have been produced with much less water and expenditure in the better water endowed areas. This is a rather wasteful way of looking at hydrological equity within the country. In place of transferring huge volumes of water over a long distance through a very expensive process, the transportation of 'virtual water' to water-scarce areas in the form of foodgrain would be cheaper. If at all the transfer of water from 'surplus' basins to 'deficit' basins has to be undertaken for such commercial purposes, it has to be seen from a different policy perspective. A pricing mechanism for payment to the donor river basins should be in place. In this way, interbasin water markets can flourish, which will ensure a more efficient utilisation of water. If such a market gets established, it may be the beginning of mutually agreed interbasin transfers between the states, providing a new way to look at interbasin transfers.

Production of foodgrains should be protected from the variations in climate. For this, however, the interlinking project is not the best available option. The better options are related to more fundamental changes in agriculture by addressing many other factors, in particular those of sustainability. Otherwise, as Postel (1999) has cautioned:

> *It is not enough to meet a short-term goal of feeding the global population. If we do so by consuming so much land and water that ecosystems cease to function, we will have, not a claim to victory, but a recipe for economic and social decline.*

However, there are no reasons to take a religious stand that interbasin transfers should not be taken recourse to. Such transfers may be taken up for rapid economic growth, which is not partial in favour of irrigated areas while creating millions of development and environmental refugees. If such transfer projects are to be taken up, several serious questions will have to be addressed first. They need to be answered before going ahead with any project. Many of these questions relate in part to the lack of an open and transdisciplinary knowledge base for even making the technical designs, and in part to the social and economic conflicts that have traditionally been inherent in such projects. As and when the project starts to be implemented, if attempts are made to avoid answering these questions, popular opposition will be the result.

Knowledge Gap in Himalayan Component

The essence of the proposed interlinking of rivers is that with the construction of storage dams as proposed, the severity of floods and the extent of flood damage will be drastically reduced. When transferred to other basins with lower water endowment, the water thus stored would reduce the regional imbalance in the availability of water in the country.

Construction of dams on the Himalayan rivers as a component of the proposed interlinking of rivers cannot be undertaken by ignoring vital questions on the uncertainty associated with taking a mechanical and traditional view of development of the Himalayan rivers (Ives and Messerli, 1989). The approximations and assumptions inherent in the standardised mathematical models of hydraulic engineering, have so far slighted and undermined the basic dynamics of sediment generation, discharge and deposition characteristic of the Himalayan rivers, which carry among the highest sediment loads in the world (Bandyopadhyay and Gyawali, 1994). Due to the verticality and the consequent fragility of the Himalaya, large dams on Himalayan rivers will be subject to high levels of seismic hazards, since the potential for earthquakes at the plate boundary all along the Himalayan foothills is well known and widely accepted (Khattri, 1987). The knowledge base required for making a professionally comprehensive assessment of such projects is still in its infancy. To any professional informed of the complexity of the eco-hydrology of the Himalayan rivers, it is clear that the development of systematic knowledge needed for making a credible impact assessment of the proposed dams and canals would need extensive field observations spread over decades. This needs to be seen in the background of the time-span of 12 years given by the Supreme Court as the time limit for the completion of the proposed interlinking.

When such a comprehensive knowledge base gets generated in an open and professional manner, in all probability several of the proposed projects may prove to be technically and economically unfeasible. Recognising the seriousness of the gaps in knowledge, NCIWRDP (1999a:187-88) took the wise view that:

> *The Himalayan component would require more detailed study using systems analysis techniques. Actual implementation is unlikely to be undertaken in the immediate coming decades.*

One example of the significance of the knowledge gap is related to the declared benefits of 'flood control' from the interlinking project. Floods in the Himalayan foothills and adjoining plains are the result of a complex ecological process involving movement of enormous volumes of water and solids. Simplistic engineering claims about projects that will control floods in Himalayan rivers are not new, and have been made over decades. Efforts to control floods in the Himalayan rivers have resulted in changes in the form of the floods. Loss from floods has not declined over the years. There is no clear scientific evidence of the ability of the proposed dams in controlling floods in the Himalayan rivers. In the background of the inadequate knowledge base, the newer interventions may be counter-productive.

IV. Interlinking of Rivers and Water Conflicts in South Asia

Even as the country receives a preordained quantity of water either through precipitation or as inflows from upstream countries, experts have been warning India of difficult times with regard to water. News of widespread conflicts has been received from many regions. Whether it is between Haryana and Punjab or Karnataka and Tamil Nadu, India has seen several conflicts related to sharing of river waters. The present crisis, however, is a result of factors that have been operating over a long time. One crucial component behind this has been the inherent failure of policy-makers to perceive the dimensions of the upstream-downstream linkages. Such gaps have led to a major source of conflict over water use in river basins all over the world.

Conventionally, the thinking has been that inequitable distribution of fresh water leads to violent inter- and intra-basin conflicts. The statement made by president Abdul Kalam ('Interlinking of Rivers to the Driver of Growth: Kalam', *Business Line,* February 23, 2003) widely supports this view on the basis of which he has said, "The plan to integrate the rivers of India (through interlinking) will be a key driver of the growth of the country and it would not only bring about economic prosperity but emotional integration as well." Mitra (2003) tries to find a justification to this argument:

> *Whilst we have failed in the course of more than half a century to resolve amicably the intra-basin quarrels, it will be sheer lunacy if, on top of that, a more contentious issue, that of interbasin water equity, is added to the already confused picture.*

Water conflicts are frequently generated not by an inherent scarcity in a region, but over the sharing of additional supplies. The long-drawn Cauvery dispute is a case in point. This is a conflict of interest between a downstream state (Tamil Nadu) that has a long history of irrigated agriculture and an upstream state (Karnataka) that was a late starter in irrigation development (but has been making rapid progress). The situation with respect to the SYL canal in the north is no different. In all such cases, factors like end-use efficiency and sustainability of irrigation practices have given way to war over the quantity of water. In the present context of interlinking, however, it can be safely said that unscientific and wasteful uses of water cannot provide hydrological equity in a country.

Bandyopadhyay and Perveen (2003) have argued that the proposed interlinking of rivers has the potential for generating four distinct types of conflicts. These are:

- Compensation for resettlement and rehabilitation of the displaced.
- Compensation for environmental damages from the project.
- Sharing the benefits and costs of the project among the states.
- Cooperative management of the project in international river basins.

While the traditionally followed concept of arithmetical hydrology will be able to calculate the supply-oriented requirements and project them in both space and time, the problem will arise on the requirements that arithmetical hydrology fails to recognise. Water being a state subject, the issue of transference of riparian rights under the centralised river link proposal needs to be resolved simultaneously. Though politicians have been talking of the need for shifting control over the water sector from the states to the centre, at least partially, no clear thought has been given on how to alter the present institutional structure to accommodate these modifications amicably. This is one topic that is immensely important and at the same time complex.

Inevitably, interstate trans-boundary water-related conflicts may become inter-country. Iyer warns that the proposal risks major confrontation with Bangladesh, which receives much of its water from the Ganga and the Brahmaputra, after they leave India. The water resource minister of Bangladesh has reportedly said that his government had protested to India but had so far not received any response (Vidal 2003). In effect of which, it has stated that:

> *The north of Bangladesh is already drying out after the Ganges was dammed by India in 1976. Now India is planning to do the same on (many of) the 53 other rivers that enter the country via India. Bangladesh depends completely on water... We want no kind of war, but international law on sharing water is unsure and we would request the UN to frame a new law. It would be a last resort.*

From this arises the challenge to organise water related administrations so that the river basin becomes the focus of activity and the various (competing) water use sectors have an appropriate voice (Helming and Kuylenstierna, 2001).

Evidently, so far, of the innumerable conflicts that have taken place over water, many have been over quantity and infrastructure. Unfortunately, due to inadequacy of data, many plans for river basin development in developing countries are inflexible and rarely provide alternative strategies. And although more than 300 treaties have been signed by countries to deal with specific concerns about international water resources and more than 2,000 treaties have provisions related to water, countries have not devoted funding to manage surface and sub-surface water jointly, scientific data is not freely shared and the requisite spirit of cooperation is often lacking (Serageldin, 1995). The results are that damages to the ecosystem services or losses in downstream population have been ignored. Such a situation applies also to another major project for interbasin transfer, the South to North Water Transfer Project (SNWTP) of China, as seen in the analysis of Berkoff (2003). In the case of the proposed interlinking of rivers in India, the situation is more complex, with the downstream areas belonging to another country. It becomes necessary then to examine whether the interlinking project would end up intensifying the already bitter trans-boundary conflict over water sharing and availability from the village to the country levels.

V. Conclusions

The proposed interlinking of rivers appears more to be a project for an ad hoc transfer of water looking for suitable justifications. It cannot guarantee the security of domestic water supplies to the drier areas of India, in particular, the dry uplands. The only dependable solution to this problem lies in local-level harvesting and conservation of rainfall. If there is free competition, the proposed interlinking will not be able to compete with emerging technological solutions like desalinisation, as far as domestic water supply in the coastal belt is concerned. The projections of foodgrain requirements till 2050, used to justify the project, are exclusively based on unpublished work and are in conflict with results published by other researchers. Foodgrain production may be improved more economically by increasing end-use efficiency in the presently irrigation potential and quicker attainment of yield levels that have already been reached in experimental farms. For providing domestic water security in dry areas and to metropolitan regions, inter basin transfers may be resorted to, if found to be the best option after a comprehensive comparative assessment. That needs to be based on an approach very different from the present one of the NWDA, dominated by objectives of physical expansion of irrigation, whether needed or otherwise. Hydraulic equity at the national level does not mean undertaking projects for transfer of water at public expense, from better water-endowed river basins to dry areas for inefficient and commercial use through socially and economically wasteful projects not approved through open professional assessments. Thus, while interbasin transfer is not being ruled out in principle, the manner in which the proposed interlinking of rivers has been put forward is unprofessional. In the absence of an open professional assessment of the proposals right from the stage of pre-feasibility studies, it will be unacceptable on social, economic and ecological grounds.

(Jayanta Bandyopadhyay is a Professor at IIM, Calcutta in Centre for Development and Environmental Policy. He did his PhD from IIT Kanpur. He can be reached at jayanta@iimcal.ac.in

Shama Perveen is a doctoral student at College of Arts and Sciences at University of South Carolina. Her research area is Water Resources).

References

APO (1996): 'Changing Dietary Intake and Food Consumption in the Asia and the Pacific', report of the APO Symposium, December 1994 (Tokyo, Asian Productivity Organisation).

Bandyopadhyay J (1989): 'Riskful Confusion of Drought and Man-induced Water Scarcity', *AMBIO*, 18(5): 284-92.

– (2003): 'And Quiet Flows the River Project', *The Hindu Business Line*, Chennai, March 14.

– (2004): 'Adoption of a New and Holistic Paradigm is a Pre-Condition for Integrated Water Management in India' in G Saha (ed), *Water Security and Management of Water Resources*, National Atlas and Thematic Mapping Organisation, Kolkata, forthcoming.

Bandyopadhyay J, B Mallik, M Mandal and S Perveen (2002): 'Dams and Development Report on a Policy Dialogue', *Economic and Political Weekly*, 37(4): 4108-12, October.

Bandyopadhyay J and D Gyawali (1994): 'Himalayan Water Resources: Ecological and Political Aspects of Management', *Mountain Research and Development*, 14 (1):1-24.

Bandyopadhyay J and S Perveen (2003): 'River Interlinking in India: Questions on Scientific, Economic and Environmental Dimensions of the Proposal' in A K Ghosh, P K Sikdar and A K Dutta (eds), *Interlinking of Indian Rivers*, ACB Publications, Kolkata.

Berkoff J (2003): 'China: The South-North Water Transfer Project – Is It Justified?', *Water Policy*, 5(1):1-28.

Biswas A K (1976): *Systems Approach to Water Management*, McGraw-Hill, New York.

Businessworld (2001): 'Chasing the Monsoon, Finding a Drought', January 1, *<http://www.businessworldindia.com/archive/010101/cover4.htm>*

Carruthers I and J Morrison (1994): '2020 Vision – Dramatic Changes in the World Agricultural and Industrial Production Systems', *IIMI Review*, 8 (1):14-20.

Cosgrove W and F R Rijsberman (2000): *World Water Vision: Making Water Everybody's Business*, Earthscan, London.

Costanza R *et al.* (1997): 'The Value of Ecosystem Services', *Nature*, 38:253-60.

CWC (1998): *Water and Related Statistics*, Central Water Commission, New Delhi.

Deursen W P A (2000): *Humans, Water and Climate Change,* The Dutch National Research Programme on Global Air Pollution and Climate Change, Carthago Consultancy, Rotterdam.

Dyson T (2001): 'The Preliminary Demography of the 2001 Census of India', *Population and Development Review*, 27(2):341-56, June.

Falkenmark, M *et al.* (1980): *Water and Society: Conflicts in Development*, Part 2, Pergamon Press, Oxford.

Falkenmark M and C Folke (2002): 'The Ethics of Socio-ecohydrological Catchment Management', *Hydrology and Earth System Sciences*, 6(1):1-9.

Falkenmark M, C Folke and L Gordon (2000): 'Water in the Landscape: Functions and Values' in J Lundqvist *et al.* (eds), *New Dimensions in Water Security*, FAO, Rome.

Gazmuri R (1992): 'Chilean Water Policy Experience', paper presented at the *Ninth Annual Irrigation and Drainage Seminar,* Agriculture and Water Resources Department, The World Bank, Washington, DC.

Gleick P H (1996): 'Basic Water Requirement for Human Activities: Meeting Basic Needs' in *Water International*, 21:83-92.

– (1998): *The World's Water 1998-1999: The Biennial Report on Freshwater Resources*, Island Press, Washington, DC.

– (1999): 'A Human Right to Water', *Water Policy*, 1(5):487-503.

– (2000): *The World's Water 2000-2001: The Biennial Report on Freshwater Resources*, Island Press, Washington, DC.

Goyal P (2002): 'Food Security in India', *The Hindu*, online edition, January 10, *www.hinduonnet.com/thehindu/biz/2002/01/10/stories/ 2002011000440200. htm*

Hanchate, A and T Dyson (2004): 'Prospects for Food Demand and Supply', in T Dyson, R Cassen and L Visaria (eds), *Twenty-first Century India: Population, Economy, Human Development and the Environment*, Oxford University Press, New Delhi.

Helming S and J Kuylenstierna (2001): 'Water – A Key to Sustainable Development', Issue paper for the International Conference on Fresh water, Bonn, December 03-07.

Ives J D and B Messerli (1989): *The Himalayan Dilemma*, Routledge, London.

IWRS (1996): 'Interbasin Transfers of Water for National Development: Problems and Perspectives', Indian Water Resources Society.

Iyer, R R (2000): *Water: Charting a Course for the Future*, Centre for Policy Research, New Delhi.

Khattri K N (1987): 'Great Earthquake Seismicity Gaps and Potential for Earthquake Disaster Along the Himalayan Plate Boundary', *Techno-physics*, 138:79-82.

McCaffrey, S C (1992): 'A Human Right to Water: Domestic and International Implications' in *Georgetown International Environmental Law Review*, 1:1-24.

Mitra A (2003): 'Knotty or Just Nutty? – A Unified River System Is Likely, to Be a Mirage', *The Telegraph*, Kolkata, May 30, *http://www.telegraphindia.com/1030530/asp/opinion/story_2016948.asp*

MoA (2003): *Annual Report 2002-2003*, Department of Agriculture and Cooperation, Ministry of Agriculture, New Delhi, *http://agricoop.nic.in/ cropprod02.htm*

Mohile A D (1998): 'India's Water and Its Plausible Balance in Distant Future', unpublished paper as cited in NCIWRDP, (1999a:29).

NCIWRDP (1999a): Integrated Water Resource Development: A Plan for Action, National Commission on Integrated Water Resource Development Plan, MOWR, New Delhi.

– (1999b): Report of the Working Group on Interbasin Transfer of Water.

Nigam A, B Gujja J Bandyopadhyay and R Talbot (1997): Freshwater for India's Children and Nature, WWF and UNICEF, New Delhi.

Postel S (1999): Pillar of Sand, Can the Irrigation Miracle Last?, WW Norton, New York.

Prabhu, Suresh P (2003): 'Interview' in *Indian Express*, New Delhi, March 2.

– (2004): 'Garland of Hope: River-Linking as a Solution to Water Crisis', *Times of India*, Kolkata, August 14.

Ravi C (1998): Food Demand Projections for India, unpublished study (as. quoted in NCIWRDP, 1999a:51).

Serageldin I (1995): *Toward Sustainable Management of Water Resources*, The World Bank, Washington, DC.

Shiklomanov I A (1993): 'World Freshwater Resources' in P H Gleick (eds), *Water in Crisis*, Oxford University Press, New York, pp 13.

Singh B (2003): 'A Big Dream of Little Logic', *The Hindustan Times*, New Delhi, March 9.

Swaminathan M S (1999): *A Century of Hope: Harmony with Nature and Freedom from Hunger*, East West Books, Madras.

TFILR (2003): 'Interbasin Water Transfer Proposals', Task Force on Interlinking of Rivers, Ministry of Water Resources, GoI, NewDelhi.

Vaidyanathan A (2003): 'Interlinking of Rivers', *The Hindu*, Chennai, March 26.

Verghese B G (2003): 'Waiting for Godaganga', *Outlook Magazine*, June 30.

Vidal J (2003): 'India's Dream Bangladesh's Disaster', *The Guardian*, July 24, *http://www.counter currents.org/en-vidal240703.htm*

Wang, Shucheng (2002): Resource Oriented Water Management, China WaterPower Press, Beijing.

Wittfogel K A (1957): Oriental Despotism: A Comparative Study of Total Power, Yale University Press, New Haven.

WMO (1992): 'The Dublin Statement', World Meteorological Organisation, Geneva.

Wolff G and P Gleick (2002): 'The Soft Path for Water' in P Gleick (ed), *The World's Water 2002-2003*, Island Press, Washington.

World Bank (1999): India: Water Resource Management – The Irrigation Sector, Allied Publishers, New Delhi.

World Commission on Dams (2000): Dams and Development – A New Framework for Decision-Making, Earthscan Publishing, London. *www.dams.org/docs/report/wcdreport.pdf*

Latest on Interlinking of Rivers

President APJ Abdul Kalam is in favour of Interlinking River Project (ILR). On 24th September 2006, while addressing the silver jubilee celebrations of Tamil University at Thanjavur, he claimed that it will be completed in 10-15 years. There are many other supporters to this project – Rajnikanth, Karunanidhi, Jayalalitha, Subramaniam Swami, Atal Behari Vajpayee, despite of the fact that the states of Kerela, West Bengal, Bihar and Punjab are against this project.

Ministry of Water Resources, Government of India has set up the National Commission for Integrated Water Resource Development (NCIWRD). It says that the river linking data of Himalayan region is not freely available and with the help of public information it seems that river linking component is not feasible in Himalayan region. About Peninsular river linking component, NCIWRD concludes that massive water transfer is not important. The only thing required is the efficient utilization of intra-basin resources. Cauvery and Vagai basins are facing deficit and thus must be taken care while diverting water from Godavari.

According to the information given by the Minister of State for Water Resources, Jai Prakash Narayan Yadav, National Perspective Plan (NPP) for Water Resources Development has identified thirty links. Out of these, feasibility report for 16 links has been prepared by the National Water Development Agency (NWDA).

A group was formed in June 2002, consisting of Central Water Commission (CWC) officers, NWDA officers, Secretaries of Irrigation/Water Resources Departments of different states involved, which was headed by Chairman of CWC. The objective of this group was to come to a consensus after discussing the water sharing issue with the concerned states. This group is working on five selected peninsular links and will try to reach a consensus among the states. Government of India and state governments of Madhya Pradesh and Uttar Pradesh signed a tripartite agreement on 25th August 2005, after which Detailed Project Report (DPR) for Ken-Betwa link has been taken up by NWDA. For the Parbati-Kalisindh-Chambal link, states of MP and Rajasthan have approved the preparation of DPR.

Compiled by Shashikala choudhary, Research Associate, Icfai Business School Research Centre, Ahmedabad.

4

Irrigation within Integrated Catchment Management

Geoff McLeod and Scott Keyworth

Over $40m have been invested in some 70 individual projects as part of the Murray-Darling Basin Commission's Irrigation Regions Program since 1992. The more recent focus of this investment has been towards the strategic development of sustainable irrigation practices within an Integrated Catchment Management context. A number of frameworks using a risk management approach were developed to support future public investment in irrigated agriculture, including approving irrigation development proposals, optimizing water use, managing water quality and investment to improve water use efficiency.

Targeted catchment plans and effective partnerships between catchment managers and industry are critical to achieving environmental objectives across geographic scales and meeting market requirements. Clarity of vision for the irrigation sector, objectives and strategies that can be translated to farm level activity, priority actions developed in partnership with industry and efficient information

management and reporting are all critical elements in catchment plans to guide farm level action in a way that will achieve the required outcomes at catchment, basin and national scales.

The Commission's studies found that the catchment plans must be supported by policies that facilitate investment to improve water management and institutional arrangements that account for the interconnectedness of the hydrologic systems present across the Murray-Darling Basin. This improved catchment planning will provide significant cost savings and benefits to industry, community and governments.

Introduction

The Murray-Darling Basin Commission (MDBC) has invested, as part of its Strategic Investigations and Education program,. over $40m in 70 individual projects as part of an Irrigated Regions Program since 1992. This Program undertook activities to produce practical information to support resource managers, policy development, regional planning and implementation of on-ground works.

The most recent suite of projects was initiated in 2002, under a Watermark banner. These projects were built upon the findings of the earlier work, addressing strategic development of sustainable irrigation practices in the context of the commitment made by the MDBC Ministerial Council to Integrated Catchment Management (ICM). It was expected that many of the findings of this work would also be relevant to irrigation regions outside of the Basin.

The findings of these projects have recently been synthesised within a series of theme reports in relation to the current issues facing irrigation within the Murray- Darling Basin. These theme reports address issues such as farming systems and best management practices, alternative approaches to adoption of better practices, improved water management, channel seepage, ground water use and management, environmental stewardship, information management and biodiversity.

Key Findings

A number of key findings have resulted from the investment in these most recent projects. Following is an outline of some of the more significant findings.

Water Use Efficiency (WUE)

Within irrigation, WUE has many complex technical and socioeconomic dimensions. The focus of the study (Capital Ag 2003, 2005) was to develop a holistic policy framework for public and private investment to improve productivity growth and environmental outcomes through improved water management within the Basin.

The study found there is significant opportunity to improve water use efficiency within the MDB with potential savings being in the order of 900-3000 Gl. Currently, the overall level of farm WUE in the MDB, however, continues to improve at about one percent per annum.

The WUE policy agenda is about the investment environment in which farmers and irrigation water providers operate. Adoption of technology by irrigation farmers and water authorities is strongly influenced by where an individual is within their investment cycle. At all points in an investment life cycle there are critical factors that influence the farmer's willingness to commit to investment in WUE. As such, Government and industry policy to improve water use efficiency is in reality a investment policy for irrigators.

An investment policy framework has been proposed that provides a strategic focus on policies, legislation, measures that influence investment levels and mechanisms to inform both investors and those who influence the investment processes.

Environmental Stewardship

There is ongoing public pressure for Australian irrigated agriculture to improve its environmental performance. It has also been anticipated for some time that markets will demand that farmers provide assurance that production is occurring using environmentally sustainable practices. Environmental Management Systems have been promoted as a means to provide this assurance. However, to date, these systems do not guarantee environmental performance outcomes that are consistent with catchment requirements.

The study (URS 2005) found that current arrangements for natural resource management delivery have four major deficiencies: actions required at a catchment scale to achieve basin scale targets are poorly defined; actions required at the farm scale to achieve catchment targets are poorly defined; reporting of outcomes against catchment and natural resource management targets is not well aligned across geographic scales and environmental quality assurance schemes and associated auditing arrangements are not integrated.

Deconstructing the environmental management system requirements by separating the audit process from environmental performance component provided the basis for developing an Environmental Stewardship System (ESS) that embodied continuous environmental improvement linked to environmental and catchment targets.

The ESS provides a framework for aligning and clarifying objectives and targets across geographic scales and an auditable system to recognise land managers who deliver environmental stewardship. The study highlighted that the focus for most farmers should be on improving environmental performance rather than assurance processes because this is what delivers catchment and other environmental outcomes. The appropriate rigour of the assurance processes is dictated by the strength of external drivers (market or regulatory) to provide the benefits to justify the audit costs. Currently these external drivers are not strong for most agricultural industries.

Water Quality

Irrigated agriculture creates an increased risk of off-site contaminant transport due to the movement and application of water, the disturbance of soils and the proximity of most irrigation areas to river systems. The actual impact of irrigation activity is dependant upon a range of factors, including the degree of connectedness of streams with a region's water delivery infrastructure and groundwater system.

The study (URS 2004) highlighted that there is limited information available to assess the relative impacts of irrigation areas on catchment scale water quality. Current monitoring systems are generally only useful for assessing the aggregate impact of all upstream effects of a limited number of variables.

In recognition that the spatial and temporal impacts of irrigation are highly variable, greater emphasis needs to be given to water quality and ecological condition in catchment water management plans. Modelling and field verification are considered key tools to support on-ground actions given the diffused nature of agricultural inputs and variable seasonal conditions.

Water quality targets set by catchment and water management authorities are rarely defined with specific water quality outcomes or time frames for achievement. This is changing with the establishment of end of valley targets (e.g., salinity) and imposition of license conditions for the discharge of drainage water from some areas.

The study found that irrigators are willing to play their part in addressing end-of-valley water quality targets, provided the need for action is identified convincingly. There are few economic drivers for irrigators to improve their land and water management practices in order to reduce their water quality impacts on downstream communities. The study concluded that each of the industries examined had the capacity to generate improved water quality outcomes provided that the water quality issues are clear, relevant to the industry and changes in management will make quantifiable improvements.

This study developed and trialed a risk based investment planning framework based upon the use of natural resource economics to provide a platform for trade-off between economic, social and environmental outcomes to achieve improved water quality. The current lack of biophysical modeling linking changes in land and water management with water quality and ecological outcomes is a limiting factor. However, the framework is considered sufficiently flexible to add value to the future adaptive management of water quality.

Groundwater Use

There has been a significant increase in groundwater use since 1999-2000 as a result of increased agricultural development and reduced surface water availability. Increased groundwater extraction in catchments with productive aquifers that are hydraulically connected to adjacent streams, creates significant potential for the reduction in stream flow to occur. It has been estimated that the recent increase in groundwater extraction will reduce stream flow within the MDB by some 550 Gl each year (REM 2003).

A preliminary examination of the drivers of groundwater use, highlighted that there was no single factor controlling the use of this resource; biophysical, social, economic and regulation all play a role across individual Groundwater Management Units (GMUs) (REM 2005). However, it was concluded that policy and regulation are the only factors that can limit groundwater use to sustainable levels.

Groundwater models developed as a part of this study showed that the onset of the initial impact of groundwater pumping on stream flow is rapid and is long lived (typically 20 or more years). Therefore, management arrangements should assume that all groundwater extracted will eventually impact upon a connected river.

Groundwater and surface water continue to be managed as separate water resources within most regions of the MDB. The study recommended a move towards management of the total water resource through the development of a new planning framework.

In the short term, intervention through policy and regulation is the only option available to ensure that groundwater use remains sustainable, and that impacts on streamflow are maintained at acceptable levels. A management approach targeted at the specific requirements and nature of each GMU is required.

Information Management and Reporting

A major weakness to developing informed surface and groundwater policy responses is the continuing lack of reliable, consistent information on irrigation including water resource availability, its use and the consequences. A study by SKM (2002) highlighted that the available information does not answer, with confidence, the most basic irrigation related questions. The study confirmed, that while there are a large number of systems and data sets available from a range of organizations, the data could not be readily accessed or consistently represented to accurately characterize irrigation at a catchment scale. Most investment made in monitoring and reporting is not outcome focused, or suitable for informing the future management of resources.

The second phase of this project conducted four case studies across a range of catchment organizations, water agencies and industries to test and refine a proposed Irrigation Management Information Reporting System (IMIRS). The study proposed an inter-operable system, at a catchment level, to enable effective reporting across catchment and basin scales. An interoperable system enables data to be collected and stored at different locations, in accordance with particular standards and accessed via the web to prepare a range of different reports.

IMIRS consists of four generic components – an information framework requiring primary data sets to allow stakeholders to generate customised local reports and a standard set of ICM basin reporting templates; technical architecture for an internet based distributed network of interoperable data sets, products and services; a cultural change component to address return on investment, connections between farm and catchment scale impacts and fear of misinterpretation of information, and; an organizational structure including formal agreements between parties covering aspects of data and information management and reporting.

The system would report on key social, economic and environmental indicators of current status, risks, investment and policy effectiveness and progressive outcomes.

Land Use and Catchment Planning

A study (SMEC 2002) was undertaken to review natural resource planning and implementation processes at a sub-catchment level. The approach of the 22 land and water management plans being implemented at that time was reviewed. The more successful plans were characterized by; presence of strong drivers particularly motivation, clarity of vision, capacity of the working group and level of community and government support; involvement of all relevant stakeholders including equitable distribution of costs and benefits; operational factors including project management, timeliness and planning processes; and content and scope including a strategic framework, sound knowledge base, scope of the issues and detailed works and measures.

A study undertaken by SKM (2004) concluded that the planning assessment and approval process for both new irrigation development and redevelopment of existing irrigation areas within the MDB is fragmented and reflects the differing histories of water and land management approaches within each State.

It was concluded by these studies that the desired outcomes expressed in the many catchment strategies and action plans in the MDB will not be achieved unless clear links are established between activities within the catchment and catchment objectives. It is imperative that sub-catchment plans continue to have an effective review process that provides for ongoing improvement and evaluation against evolving catchment objectives. Therefore, it is also essential to assess any development proposal in the context of its contribution to desired catchment outcomes.

A framework and explanatory guidelines for the assessment of irrigation development proposals were developed. The guidelines define the elements of the Land Use Suitability and Capability (LUSAC) framework and aim to inform proponents and authorities of the information provided in the development proposal and help them to understand the assessment process.

The key features of the framework included – development of a vision for irrigation within the catchment, establishment of transparent technical criteria and thresholds, preparation and submission of a development proposal, assessment and decision process to review the merit of the proposal against the criteria and thresholds, performance accountability of the developer driven by consent conditions, and a review process involving audit, monitoring and reporting to ensure that the consent conditions are implemented by the developer and that the consent authority is consistently applying the process to all applications.

Discussion

A key focus for the irrigation industry in Australia is to improve natural resource management at the farm and catchment level in a profitable manner. The irrigation sector has demonstrated the capacity to change where the requirements for change are clear and justifiable.

To improve natural resource management at the farm and catchment scale consistent with the outcomes agreed to by government at a national level requires a clear vision for irrigation at a catchment scale, greater clarity of action at a catchment and farm level and the establishment of strong partnerships between catchment organisations and industry. Transparent criteria and thresholds are also required to guide development and redevelopment of irrigation and for the development and review of sub catchment land and water management planning.

Developing targeted catchment plans that will achieve environmental objectives across geographic scales will, in some situations, need to be preceded by development of further knowledge to understand the relationship between farm actions and catchment level impacts. For example, the water quality study highlighted the need to understand the contributions of irrigation activities on concentrations and loads of relevant water quality variables. In other situations this will require further clarification of catchment targets including recommendations at a spatial scale.

The use of a risk assessment approach at a catchment, industry and farm level underpinned by objective and efficient audit and review processes is necessary to accommodate the incompleteness of knowledge of the biophysical issues and the uncertainty surrounding economic issues related to irrigation.

The alignment of reporting of outcomes against natural resource management targets across scales is critical to enable an effective review process to inform the evolution of catchment objectives and targets and the modification of strategies and approaches for future management arrangements. The complexity of economic, environmental and social issues being faced by irrigation requires high quality up-to-date information across various scales. The existing information management arrangements do not provide for this.

Strong and effective partnerships are required between industry and catchment management to provide the clarity of action required at the farm level to achieve the dual outcomes of agreed natural resource management performance and satisfying market requirements in a manner profitable to the farmer.

Given that a significant level of investment is required to improve water use efficiency and the rate of adoption will be strongly influenced by individual's capacity to invest, the development of stronger partnerships between finance providers and individuals or organisations will be necessary to enhance the level of investment. An individual's capacity to invest is influenced by where they sit within the normal investment cycle, what obstacles exist and what capacity is needed to reduce the risk of the future investment.

The risk-based frameworks and approaches developed by the Irrigation Regions Program for land use planning, environmental stewardship, water quality, groundwater management, information management and for decision-making at a community level address these key issues. Implementation will be dependant on the presence of strong drivers including motivation, having people with the skills and capacity and the availability of financial resources both at a public and private level.

Next Steps

An overwhelming message from these studies is the need to get catchment planning right. This includes establishing clarity of purpose and translation of requirements across the geographic scales, in a manner that is cognisant of market requirements. The focus of next phase of investment should include:

At a Policy Level

- Recognize the importance and complexities of investment and adopt a more coordinated approach to policy development and implementation within the context of an investment policy framework.
- Adopt a precautionary approach to water management where groundwater and surface water systems are connected. In the shorter term intervention is required to minimize creation of longer term impacts on stream flow from groundwater extraction and in the longer term develop a planning framework that manages the total water resource in an integrated manner.
- Develop an information management reporting system for irrigation land and water management as an integral part of the national natural resource management framework, capable of providing the basis for review and evaluation of sub catchment and catchment plan implementation outcomes.

At an Operational Level

- Implement the environmental stewardship system in situations where there is the combination of willing catchments and industries. By addressing the weaknesses identified in existing catchment planning and implementation processes, guidance will be provided for a future national framework for natural resource management.

(Geoff McLeod has been contracted by the Murray Darling Basin Commission for the past 4 years to coordinate the investment projects undertaken as part of its Irrigated

Regions Program. Prior to this role, he was the Environmental Manager with Murray Irrigation Limited for 6 years, having completed a Bachelor of Agricultural Science and worked with the NSW and Victorian Departments of Agriculture for 13 years.

Scott Keyworth is Manager, Research Adoption, with the Water for a Healthy Country Flagship. Prior to joining CSIRO, he was Director, Strategy Implementation for the Murray-Darling Basin Commission.)

References

Capital Agricultural Consultants (2003): Policy Frameworks for Water Use Efficiency in the Murray Darling Basin: Principles, Concepts, Approaches and Tools I2115, MDBC.

Capital Agricultural Consultants (2005): An Investment Policy Framework to Deliver Productivity Growth and Improved Environmental Outcomes Project I3005, MDBC.

REM (2003): Watermark Sustainable Groundwater Use within Irrigated Regions Stage 1 Final Report I2113, MDBC.

SKM (2002): The Development of a management Information an Reporting System for Irrigation in the MDB – Stage 1 Final report, MDBC.

SKM (2004): Guidelines for Land use suitability and capability for irrigation planning and development Final Report, MDBC.

SMEC (2002): Review of natural resource planning and implementation processes in selected irrigation regions throughout Australia Main Report, MDBC.

URS Australia (2004): Water quality in irrigated areas final report I2112, MDBC.

URS Australia (2005): Watermark environmental stewardship project I2116, MDBC.

SECTION II

KEY ISSUES

5

To Dam or not to Dam? Five Years on from the World Commission on Dams

WWF has prepared this report analyzing the impact of recommendations in WCD's report launched five years back. WCD suggested some recommendations and strategic priorities for the construction of dams. This report includes six cases where the government has failed to comply with WCD recommendations. It mentions some very critical problems related to river exploitation. WWF found that WCD report has proved to be important to reduce environmental and social challenges. To make this effort more successful, United Nations Dams and Development Project (DDP) follow up the work of WCD.

Introduction

On 16th November 2000, Nelson Mandela helped to launch the report of the World Commission on Dams (WCD), indicating the importance attached to the issue of dams and development by one of the world's greatest statesmen. The 380-page report addressed the benefits and impacts of dams or, in Mandela's words, 'one of the battlegrounds in the sustainable development arena'. Now,

Source: http://www.panda.org © WWF. Reprinted with permission.

five years on, as the dust has settled, we ask – what is the Commission's legacy? Are fewer bad dams being built? Are benefits being shared with affected communities and are more effective environmental protection measures being taken?

This is a pertinent time to ask these questions as dams, in particular hydropower projects, have recently risen back to the top of decision-makers' agendas. This year, the World Bank approved funding for the Nam Theun 2 hydropower project in Laos, its first major investment in this sector since the Bank announced in 2003 its intention to re-enter dams financing with a focus on 'High Reward, High Risk' projects[1]. Rising fossil fuel prices, growing energy needs, as well as the ratification of the Kyoto Protocol on climate change all have resulted in a renewed effort to develop the world's hydropower potential. At the same time climate change is likely to increase the demand for water storage. While hydropower and other dams undoubtedly have a role to play in meeting growing energy and water needs, there is also much at stake, as in the past too many projects have resulted in excessive environmental damage and negative social impacts, especially for local communities.

In this report, WWF takes stock of what has happened in the five years since the launch of the WCD report. We highlight six cases where governments and dam builders have failed to clean up their act. We also show a number of positive developments from around the world. Overall, we find that the WCD recommendations are as important today for reducing the social and environmental damage caused by dams as they were five years ago. WWF is convinced that applying the WCD's framework, adapted to individual country's situations, will result in better decision-making and projects which that have less impact. The world's ailing rivers and the communities that depend on them face a bleak future without prompt action.

What was the World Commission on Dams?

The Commission was established in 1998 as an independent, international, multi-stakeholder process to address what had become one of the most controversial areas of infrastructure development. While dams are seen by some as essential for development and poverty reduction, others claim they actually increase poverty, as well as damage ecosystems. The dam debate had become increasingly polarized during the 1990s and one of the aims of the Commission was to bridge the gulf between the two camps and produce an independent assessment of the performance of dams. Furthermore, it was charged with developing internationally accepted standards, guidelines and criteria for decision-making in the planning, design, construction, monitoring, operation and decommissioning of dams[2].

The Commission comprised 12 independent Commissioners and was chaired by Professor Kader Asmal, one of the ministers in Mandela's cabinet. It spent two and a half years assembling what was undoubtedly the most thorough assessment of dams ever. Yet even with nearly 1000 submissions and countless consultations with stakeholders, the Commission could only examine a relatively small sample of the world's 45,000 large dams[3] in detail.

What the Commission found[4] was that whilst dams have indeed made important contributions to human development, in too many cases an unacceptable and often unnecessary price has been paid, especially in social and environmental terms. In particular it highlighted the adverse impacts on an estimated 40 to 80 million people displaced by dams, on downstream communities and on the natural environment.

To ensure that dams do not impose excessive social and environmental costs, the Commission identified five Core Values that need to be applied to decision-making on water and energy development (see box 1). It went on to recommend a new framework for decision-making, based on seven Strategic Priorities, including the need to gain public acceptance, comprehensive options assessment and sharing benefits. The Commission also developed more detailed Policy Principles and Guidelines. To adapt these Guidelines to specific cases, the Commission identified five key decision stages, including a needs assessment and the evaluation of alternatives before the decision to build a dam is made. Where a dam is found to be the preferred development alternative, three more critical decision points occur in the stages of project preparation, implementation and operation.

The report received a welcome from many quarters, and there has been widespread agreement on the five Core Values. The recommendations of the WCD also reflected a growing international consensus on issues such as integrated water resource management (as adopted in the World Summit on Sustainable Development Plan of Implementation) and integrated river basin management (now a legal requirement in the European Union under the Water Framework Directive), covered under Strategic Priority4. But there has also been criticism, in particular from within industry, targeted above all at the Policy Principles and Guidelines. Not surprisingly, the dam controversy was not going to go away overnight.

New Evidence on the Scale of the Problem

Since the publication of the WCD report, a number of authoritative assessments have further revealed the scale of the impact of the world's 45,000 large dams and associated developments, in particular irrigation infrastructure. A study by Nilsson *et al.*,[5] has shown that already 59% of the world's large river systems are fragmented by dams. Overall, humans currently use 54% of accessible runoff[6] and several major rivers, including the Nile, Yellow, and Colorado Rivers, at times no longer reach the sea. The Millennium Ecosystem Assessment[7] found that the amount of water impounded behind dams quadrupled since 1960 and that three to six times as much water is held in reservoirs as in natural rivers.

In environmental terms, the effect of this water exploitation is serious. According to the Millennium Ecosystem Assessment, freshwater ecosystems tend to have the highest proportion of species threatened with extinction. The impacts of dams are not just localised. In another recent study Syvitsky *et al.*,[8] estimated that globally, reservoirs are holding over one billion tonnes of sediment, preventing sediment transport to coastal areas, reducing nutrient delivering to agricultural areas and increasing coastal erosion rates. These studies underline that controlling the adverse impacts of dam development is as urgent as ever.

No More Bad Dams?

Dam construction continues at a rapid pace, in particular in the developing world where growth of water and electricity demand is strongest. As shown in figure 1, China, Iran and Turkey lead in the construction of large dams, although industrialised Japan is not far behind. Currently, close to 400 large dams over 60 metres in height are under construction worldwide, as well as many smaller ones for which data is difficult to obtain. As construction periods are often long, many of these dams would have been started before the completion of the WCD report.

Box 1: Key Elements of the WCD

Key Recommendations of the WCD

5 Core Values – equity, sustainability, efficiency, participatory decision-making, and accountability.

7 Strategic Priorities

1. Gaining public acceptance.
2. Comprehensive options assessment.
3. Addressing existing dams.
4. Sustaining rivers and livelihoods.
5. Recognising entitlements and sharing benefits.
6. Ensuring compliance.
7. Sharing rivers for peace, development and security.

5 Key Decision Stages

1. Needs assessment.
2. Selecting alternatives.
3. Project preparation.
4. Project implementation.
5. Project operation.

Figure 1: The World's Largest Dam Building Nations

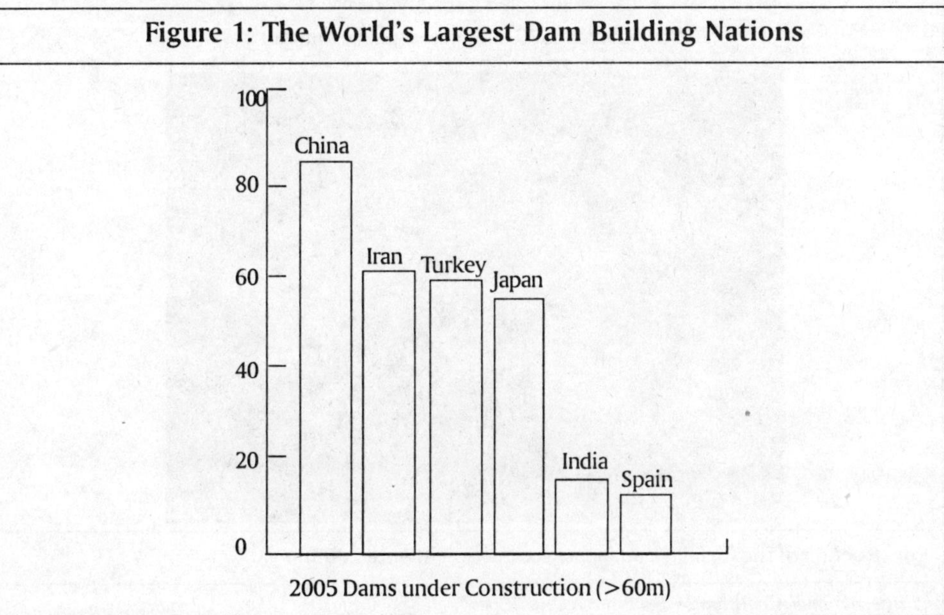

Source: International Journal on Hydropower and Dams.

It is instructive to look at some dams, which have been approved and where construction has started since November 2000, to see how they fare in terms of the WCD recommendations.

The premise of the WCD report was that a new decision-making framework would result in dams that have fewer negative impacts and greater benefits. But five years on, it is not difficult to find dams that fail to meet at least some of the recommendations of the WCD. Here we present our evaluation of six dams, which indicate that the controversy is still very much alive and that the lessons highlighted by the WCD have still not been learnt.

Case Studies

1. Chalillo, Belize

Dam height: 50 m Reservoir size: 9.53 km^2
Function: Power generation
Installed Capacity: 7 MW
Cost: US$30 million (original estimate)

Construction of the Chalillo Dam nears completion in April 2005.

© International Coalition to Save the Macal River Valley – Candy & George Gonzalez.

The decision to build the Chalillo Dam in Belize was taken in 2001 and given that the dam was going to flood more than 1000 ha of pristine rainforest, it was always likely to cause controversy. A preliminary assessment by the London Natural History Museum[10], appended as an annexe to the Environmental Impact Assessment (EIA), suggested that the dam would cause a significant and irreversible reduction of biological diversity in Belize. In particular, it threatened an already endangered population of a Scarlet macaw subspecies (Ara macao cyanoptera), possibly resulting in local extinction in Belize. The study recommended a more comprehensive and integrated long-term study of other potential sites. Unfortunately, this was not taken up by the developers, the Belize Electric Company (BECOL, owned by the Canadian company, Fortis).

While the generating capacity at the dam is relatively small, the developers cite as its main benefit, the creation of significant water storage capacity to increase power generation at a downstream hydropower plant. This is expected to increase Belize's self-sufficiency in electricity by reducing imports of power from Mexico. However the economic viability of the dam has been questioned and an increase in electricity prices was predicted[11].

Opposition to the project focused on the inadequacy of the EIA and the legal battle led by the Belize Alliance of Conservation Non-Governmental Organisations (BACONGO) went all the way to the Privy Council in London, the final court of appeal for Commonwealth countries like Belize. While the Privy Council voted by a majority of three judges to permit the project, the dissenting judgement[12] issued by the remaining two judges leaves significant room for doubt over the adequacy of the EIA.

The dam has now been completed. A large area of rainforest was cleared in preparation of flooding and there are reports of macaw nesting sites having been destroyed[13]. Habitat loss is also affecting other species, and local experts report on the decline of tapir (Tapirus bairdii) populations and reduced hunting grounds for jaguars (Pantera Onca), which already face a difficult future in Belize[14]. In addition there are reports of serious water quality problems downstream of the dam.

Following their narrow court victory, the company and the government of Belize have refused to conduct any of the legally-required follow-up on the dam's impacts or safety, and continue to refuse access to the area by independent observers[15]. Furthermore, the government has proposed another dam downstream of Chalillo, which will undoubtedly cause further ecological disturbance. In the meantime, electricity rates have indeed increased by an average of about 12%, with 14,000 households even facing rate increases of 50%[16], although the company cites rising fuel costs as the cause.

In WWF's view, this project fails to observe the WCD Strategic Priorities – 2 for comprehensive options assessment, 4 for sustaining rivers and livelihoods, and 6 on sharing benefits.

2. Ermenek, Turkey

Dam height: 210 m Reservoir size: 57.74 km²
Function: Power generation
Installed capacity: 302.4 MW
Cost: US$650 million

The site of the Ermenek Dam.

© WWF Turkey.

The Ermenek hydropower project is currently under construction on the Ermenek River, a tributary of the Göksu River in South Eastern Turkey. The Göksu River is one of the last free flowing rivers in Turkey and its delta has been recognised as a Wetland of International Importance under the Ramsar Convention. There are further plans for five hydropower plants on the main stem of the river but there has been no basin-wide assessment of the cumulative impacts of Ermenek and the other projects.

The feasibility study for the project was carried out in 1990, whereas the EIA was not carried out until 1999. No needs and options assessment has been carried out, nor have alternatives been considered such as decentralized renewables, energy efficiency or the cutting of transmission losses that are up to 30% in Turkey. Furthermore, the economic analysis carried out for the project failed to take into account various aspects such as cost of new transmission lines, losses to fisheries and future decommissioning costs[17].

There are a number of shortcomings in the EIA, which cause WWF concern about the project's eventual impacts. Ecological surveys were inadequate for assessing the potential impacts of such a large project. For example, the EIA fails to list several threatened plant species, which are known to occur in the area. The mitigation of environmental impacts was not covered comprehensively in the EIA. For instance, in WWF's view, the proposed minimum flow is inadequate to maintain downstream ecological conditions.

The project will also require the relocation of 550 people. Transparency is key to public acceptance but in the Ermenek case, cost-benefit analyses have been kept confidential and the EIA report has not been made freely available publicly due to a confidentiality agreement made between the developers and the responsible government agency, Devlet Su Iflleri (DSI-State Hydraulic Works).

In WWF's view, Ermenek falls particularly short in terms of WCD strategic priorities – 1 on public acceptance, 2 for options assessment, and 4 on sustaining rivers and livelihoods.

3. Karahnjukar, Iceland

The Karahnjukar hydropower project is being built in the highlands of Northeast Iceland, to supply electricity to a new aluminium smelter to be developed by Alcoa, the world's largest aluminium supplier. Iceland, one of the richest countries in the world, has decided to base part of its economic growth on aluminium manufacturing, even though all raw materials have to be imported from far afield. It hopes to gain competitive advantage through the supply of cheap electricity.

The project has caused both local and international controversy, with concerns about the environmental impacts of such a large project in a fragile and pristine

Dam height: 190 m (highest of 3 dams)

Reservoir size: 57 km^2

Function: Power generation

Installed capacity: 690 MW

Cost: US$1086 million

The area of the future reservoir. The sign indicates the highest level it will reach.

© *Ute Collier/WWF International.*

arctic wilderness area. More specifically, the project will flood five hundred nesting sites of the rare pink-footed goose (Anser brachyrhynchus) and Iceland's only reindeer herd is likely to diminish. Wetlands downstream will also be affected by wind erosion of soils left exposed from construction, the draining of watersheds and the fluctuations of the water level in the reservoir.

The project's EIA was at first rejected by Iceland's National Planning Agency, a decision that was later overruled by the Minister for the Environment. The project remains a divisive issue within Iceland and three Icelandic citizens and the Icelandic Nature Conservation Association took the Minister for the Environment to court for overturning the Agency's decision on the project's EIA.

A coalition of NGOs (Non-Governmental Organisations), including WWF, has brought the Kárahnjúkar case to the attention of the Bern Convention Standing Committee[18], which in 2004 issued a set of recommendations regarding mitigation measures for the project to the Icelandic Government. According to the NGO coalition, most of the key recommendations remain to be implemented.

The economics and social impacts of the project have been, and continue to be questioned. A 2005 OECD report[19] suggested that Iceland's large-scale aluminium-related investment projects might result in the overheating of the economy. The report also says that the economic returns of such projects are unclear. Furthermore, it has been suggested[20] that government support for education and eco-tourism would provide better development alternatives from an environmental and socioeconomic point of view.

While the environmental shortcomings of the project remain, one positive development has been the Icelandic government's proposal of a new national park – potentially the largest in Europe – which will protect Jökulsá á Fjöllum, an adjacent watershed. After the construction of Kárahnjúkar this will be the last free flowing glacial river in the Icelandic highlands.

In WWF's view, the Kárahnjúkar project fails to observe WCD strategic priorities – 2 for options assessment and 4 on sustaining rivers and livelihoods.

4. Nam Theun 2, Laos

Dam height: 39 m Reservoir size: 450 km²
Function: Power generation
Installed capacity: 1070 MW
Cost: US$1500 million

The Xe Bang Fai River – Water diverted from the Nam Theun 2 reservoir will affect both fisheries and river bank agriculture alongside this river.

© *Shannon Lawrence, Environmental Defense.*

In March 2005, the World Bank's decision to support the construction of the Nam Theun 2 dam in Laos propelled the project forward after decades of studies and planning since the hydropower potential of the Theun River was first recognised in the 1970's. Its developers promote the project as Laos' best chance to increase government revenues for poverty reduction through the export of electricity to Thailand. The cost of the project is estimated at nearly US$1.5 billion and it will be at least another decade before revenues will get anywhere near the US$80 million per year forecast by project proponents[21].

Of particular concern are the likely widespread social and environmental impacts, ranging from the resettlement of 5,700 villagers to the impacts on the Nakai Nam Theun Biodiversity Conservation Area and fisheries in the Xe Bang Fai watershed. At least 50,000 people who rely on the Xe Bang Fai River for their livelihoods will be affected as water is diverted from the Nam Theun River, resulting in loss of riverbank gardens, reduced access to the river, loss of fish habitat and reduction in fisheries yields. Following the publication of the WCD report and increasing international criticism of the project, an environmental and social impacts safeguards programme was put into place. Efforts to consult stakeholders were also stepped up and consultations with local communities were carried out.

Whilst the attention to environmental and social safeguards is undoubtedly more thorough than for any other project in Laos, the project has gone ahead without a proper assessment of the needs it is supposed to meet or ananalysis of whether it will be the best option to meet these needs (WCD Strategic Priority 2). Independent analysts question the forecasts of Thai energy growth used to justify the project[22]. Furthermore, throughout the region numerous other dam projects are being planned whose economic viability will depend on exports to Thailand. Even the World Bank remains cautious about Nam Theun 2's contribution to poverty reduction in Laos, stating "If the revenues are spent efficiently, accountably, and transparently – in accordance with project agreements, NT2 could provide significant, incremental support to Lao PDR's poverty eduction and biodiversity conservation efforts." (Emphasis added)[23].

The need for sustainable economic development in Laos cannot be denied, but the social, environmental and economic risks associated with Nam Theun 2 are considerable. In WWF's view, the question whether Nam Theun 2 really is the best solution to meet Laos' development needs remains unanswered. Only time will tell whether the project benefits will materialize in full.

WWF's believes that the Nam Theun 2 project fails to fully observe WCD Strategic Priority – 2 for options assessment. In particular, the project fails in the two WCD key decision stages of needs assessment and selecting of alternatives. Furthermore, there is an urgent need to take a step back from the project-by-project approach in the Mekong basin, where at least 21 more large dams are being planned, and develop a more strategic, basin-wide approach, taking into account the cumulative impacts of multiple dams within the watershed.

5. Melonares, Spain

Dam height: 42.25 m
Reservoir size: 14.57 km^2
Function: Urban water supply
Cost: US$180 million

Construction is ongoing at the Melonares Dam, despite uncertainties about how water from the dam will be transferred to Seville.

© F. Fuentelsaz WWF Spain.

Although Spain already has the largest number of large dams per inhabitant in the world, the National Hydrological Plan, approved in 2001, still included plans for the construction of another 120 dams. Several of these, Melonares, La Breña II and Arenoso, are located in the Andalusian Guadalquivir basin and will severely affect the habitat of the Iberian lynx, (Lynx pardinus), the most endangered cat species of the world.

The 1997 EIA for the Melonares Dam, currently under construction to supply drinking water for the city of Seville, highlighted the potential negative impacts of the project on the Parque Natural Sierra Norte, a Regional Nature Park as well as a Special Protection Area (Birds Directive 79/409 EEC) and Site of Community Interest (Habitats Directive 92/43 EEC). It established that the project should only be allowed to go ahead if it can be demonstrated that there are no other alternatives for meeting the water needs of Seville[24].

However, when the dam was initially approved[25] the needs assessment for urban water supply in Seville failed to take into account key factors, such as water efficiency trends. In fact Seville's water demand in 2005, a year of severe drought, is 1.34 million m^3 per year instead of the 1.75 million m^3 per year forecast in the original dam project proposal. Furthermore, a large number of alternative options, such as supply from existing dams or abstractions from aquifers, were ignored.

In this context, the potential for transfer of water allocations from agricultural to urban supply should have been considered. For Spain as a whole, a WWF analysis has revealed that four surplus crops – corn, cotton, rice and alfalfa consume the equivalent of the water requirements of 16 million domestic consumers[26]. About 88% of the water of the Guadalquivir is currently used for irrigation of often low-profit crops[27], such as maize, subsidised under the European Union Common Agricultural Policy. Spain is now implementing water markets, an option offering farmers a way to gain additional income by selling water rights. For example, the Viar Irrigators Community agreed in May 2005 to supply 100,000 m^3 per year of good quality water from their private Pintado dam to the Seville Urban Water Supply Company EMASESA. WWF believes that such water markets, together with well-designed water conservation measures would have provided a less costly solution than the Melonares dam, with fewer environmental impacts.

Nevertheless, the EU Commission – on request of the Spanish Government approved funding for the Melonares dam in October 2002. Under the Cohesion Fund the EU contributes about 85% of the budget for the dam and associated environmental mitigation measures. However, funding was suspended in 2005, as there are still uncertainties about how water will be transferred to Seville. Authorities are planning to use an existing irrigation channel, but this would open the possibility that water would be used for other purposes than the urban water supply for which the EU approved the funding.

In WWF's view, the Melonares project does not meet WCD strategic priorities 2 for comprehensive options assessment and 4 on sustaining rivers and livelihoods. In addition the project is weak in terms of WCD standards key decision stages – 3 and 4 on project planning and implementation, as incomplete plans have led to suspension of funding during the construction stage.

6. Burnett, Australia

Dam height: 37 m
Reservoir size: 29.5 km²
Function: Irrigation and domestic/industrial water supply
Cost: US$150 million

Sugar cane fields in Queensland. The Burnett dam is being built to provide irrigation water for this unprofitable agricultural crop.

While the Australian Government is spending up to US$2 billion over five years to restore water flows in the heavily dammed Murray-Darling basin, elsewhere in the country, dam building is still going ahead. In Queensland, the Burnett River Dam has been under discussion since the 1960s, as a means of improving agricultural production and encouraging urban and industrial development. It was repeatedly rejected as not viable, most recently in 1997. Nevertheless, in November

2003 construction of the dam, the first and largest of five proposed water infrastructure projects on the Burnett River, started and is today nearing completion.

There has been a lack of transparency as concerns the economic assessments of the dams. Access to most studies has been denied to the public on the basis that they contain commercially confidential information. Based on the available information there are serious concerns about the economic viability of the dam as much of it depends on the expansion of the sugar industry. However, the sugar industry has not been profitable and it is likely that producers will be unwilling to pay the high water prices that are needed to achieve full recovery of the dam's costs[28]. As a result, it is likely that the dam will require subsidies indefinitely.

In addition to economic concerns and a lack of transparency, there are also serious environmental impacts expected from the project, most notably on the Queensland lungfish, (Neoceratodus forsteri). The lungfish, listed as a nationally threatened species and protected from fishing under the Queensland Fisheries Act, is found only in a few rivers in Queensland and the Burnett River is one of its prime habitats. It is extremely specific in its choice of spawning habitat and the Burnett River Dam is likely to further reduce spawning sites and increase the risk of its extinction.

In WWF's view, the decision to build the Burnett River Dam appears to be mainly politically motivated and was taken by the Queensland Government following promises made in the lead up to the 2001 state elections. Construction of the dam was believed to bring substantial employment to the Burnett basin, both through job creation related to dam construction and through expanding agriculture. However, a least cost planning study for the Burnett region, commissioned by the Queensland Environmental Protection Agency, proposed a number of alternatives, including off stream storage and water efficiency measures, some of which would provide more jobs, cheaper water and less environmental impacts[29].

In WWF's view, this project fails to observe WCD Strategic Priorities – 1 for gaining public acceptance, 2 on comprehensive options assessment and 4 for sustaining rivers and livelihoods.

Signs of Change?

As the case studies show, there are still numerous examples of individual dam projects that fail to meet one or more of the WCD Strategic Priorities. In particular there appears to be a failure to undertake comprehensive needs and options assessments (Strategic Priority 2). The requirements of the WCD in this Strategic Priority are similar to those of Strategic Environmental Assessment (SEA) which is now being implemented in some countries but is far from common place. Furthermore, EIAs are also often inadequate. A particular problem is the lack of basin-wide assessments of cumulative impacts where multiple dams are proposed (Strategic Priority 4). Transparency, which is essential for fostering public acceptance (Strategic Priority 1), was also found to be lacking in two cases. Only one of the dams examined (Chalillo) was complete at the time of writing and in this case, it appears that benefits (Strategic Priority 5) for the local population have not materialised and people are seeing an increase, rather than the promised decrease, in electricity prices.

The six dams examined here are only a small selection of the hundreds of large dams that are under construction globally and this assessment does not claim to be comprehensive. However, the six chosen dams are indicative of a general lack of application of the WCD recommendations in key dam building countries. South Africa, Mandela's home country, is indeed one of the few countries which has embarked on a comprehensive follow up process to the WCD report. In a three-year process, a multi-stakeholder committee led the South African process to recommendations for changes in policies and procedures.

Internationally, the work of the WCD is being followed up under the auspices of the United Nations Dams and Development Project (DDP[30]). A multi-stakeholder forum with representatives from governments, industry, financing institutions, affected people and NGOs has been meeting on an annual basis for the last four years. Yet, for example in the case of Nam Theun 2, the views of the forum participants are still miles apart.

The DDP supports so-called 'dialogue' activities as a WCD follow-up and these have taken place in Argentina, Indonesia, Kenya, Namibia, Lesotho, Nepal, Malawi, Pakistan, Sri Lanka, Vietnam, Thailand and Zambia. While multi-stakeholder dialogues in these countries are to be welcomed, they have so far not yielded any specific policy changes. Some developed countries such as Germany, the Netherlands, Norway, Sweden and UK have also initiated follow-up discussions but again with few concrete results. In the case of the UK, the government produced a consultation draft response to the WCD in 2002 but never actually published the final response.

And yet, while formal and tangible responses to the WCD are few and far between, there are some signs that approaches to dam building are changing, even in countries like China, the world's largest dam building nation.

In early 2004, the Chinese Premier Wen Jiabao called for a review of the construction plans for 13 dams on the Nujiang (Salween) River, the last major free flowing river in Asia. This followed increasing public questioning of the development plans for the river, previously unheard of in China. Later in 2004, China's State Environmental Protection Administration temporarily halted the construction of 30 major construction projects that had not complied with EIA legislation. Amongst these were parts of the Three Gorges Project, the world's largest hydropower project, and the Xiluodu Dam. While construction has since resumed, this development still indicates that there is increased attention by the Chinese environmental authorities to the environmental impacts of infrastructure development.

The Yangtze is the river basin most threatened by dam building with 105 dams planned or under construction. April 2005, saw the establishment of the Yangtze Forum which brought together for the first time various central government departments, national and provincial governments, to develop a common vision for the management and conservation of the river. The Chinese authorities recognise that the Forum is essential to build relationships with key stakeholders, such as NGOs and local communities, to improve the efficiency of river management. WWF, which has worked with the Chinese government since 1980 for the conservation of the Yangtze River basin, is a catalyst and supporter of the Forum. It is hoped that the Forum will provide a model for more sustainable river management in China through enhanced integration of the activities and expertise of national and provincial agencies.

Meanwhile, the dam industry, while still broadly critical of the WCD report, and in particular the guidelines, has made some efforts to improve practice. The International Hydropower Association has published sustainability guidelines[31] aimed at promoting good practice within the industry. While not fully endorsing the WCD, the IHA stresses the acceptance of the Core Values and the objectives of the Strategic Priorities.

Furthermore, a number of financing institutions and Export Credit Agencies now use the WCD recommendations as a global reference point for the assessment of dam projects[32]. For example, the global bank HSBC Holdings, which is working with WWF in a five-year 'Investing in Nature' Partnership, published a freshwater infrastructure sector guideline in May 2005. The guideline states that HSBC will not provide facilities and other forms of financial assistance, including any involvement in debt and equity capital markets activities and advisory roles, to dams that do not conform to the WCD Framework[33].

While often the focus is on new dams, another of the Strategic Priorities of the WCD, Addressing Existing Dams (Strategic Priority 3), is increasingly receiving attention. The options for improving existing dams are numerous and can bring both environmental and economic benefits. For example, a WWF study in Brazil has suggested that the upgrading of the country's aging hydropower plants could add an extra 8,000 MW installed capacity at low cost[34]. Environmental benefits can be gained by removal of obsolete dams or by implementing environmental mitigation measures such as fish ladders. In Zambia, WWF has worked together with government and the electricity company to restore environmental flood releases from the Itezhi-tezhi dam. Here, extensive modelling work demonstrated that dam operations can be modified to improve environmental benefits, without affecting the productivity of the dam[35].

Conclusions and Calls to Action

Five years after the publication of the WCD, the debate on dams still rages. Controversy surrounds dams such as Chalillo and Nam Theun 2, and the World Bank's 'High Reward, High Risk' policy for financing large new dams. Yet, as this report has shown, there are also some positive signs of change in decision-making processes relevant to dams, if not the full-blown reform the

WCD advocated. Governments like South Africa and institutions like HSBC appear to be heeding the lessons of the past. WWF welcomes these positive developments but also deplores that some dams are still built based on dubious economic arguments, without considering all alternatives, without transparent processes and without adequately addressing serious environmental and social impacts, as demonstrated by the six case studies in this report.

On the fifth anniversary of the WCD, WWF thus urges decision-makers to revisit the findings of the WCD and revise their policies in accordance with the WCD recommendations. In particular, there is a need for dam decision-making to take place within the frameworks of Integrated Water Resources Management and Integrated River Basin Management to ensure that a balance is struck between economic, social and environmental issues within river basins. Where there are plans for several dams in the same river basin, decision-making must move away from a project-by-project approach and assess the benefits and impacts of dam development on a river basin level. From an ecological point of view the assessment of the cumulative impacts of multiple dams within a river system is of key importance. Furthermore, a more comprehensive approach to options assessment is necessary to ensure that alternatives are properly considered. As demonstrated by the case studies in this report, many projects still go ahead without proper needs and options assessment. Strategic Environmental Assessment is a tool particularly suited for delivering both options and cumulative impact assessment but is rarely implemented.

With increasing pressure to develop new dam projects, in particular in developing countries, now is the time to ensure a more systematic implementation of the WCD's recommendations. They are as important for reducing the extensive social and environmental damage caused by dams today as they were five years ago. WWF is convinced that applying the WCD's framework, adapted to individual country's situations, will result in better decision-making and projects that have less impact. The world's ailing rivers and the communities that depend on them face a bleak future without prompt action.

Governments and dam builders have already had five years to clean up their act. WWF – the global conservation organization, says that its time now for governments, dam builders and financiers to implement the WCD recommendations or face a growing public backlash from the unacceptable economic, human and environmental costs of badly planned dams. In particular, they must:

1. Assess needs and options more comprehensively, with particular attention to options for managing the demand for water and power to minimize the need for new dams.
2. Consider new dams only after strategic environmental assessment to ensure that whole river basins are sustainably managed.
3. Ensure that, wherever feasible, existing dams are retrofitted to increase power generation and other economic benefits while reducing social and environmental impacts.

The development of new dams in accordance with the seven Strategic Priorities recommended by the WCD is the best way to ensure that dams really deliver their intended benefits and avoid unacceptable impacts.

(WWF is the world's largest and most experienced independent conservation organization; conserves endangered species, protects threatened habitats and addresses global threats. It finds long-term solutions that benefit both people and nature.)

References

1. World Bank (2003), Water resources sector strategy: strategic directions for World Bank engagement, Vol.1 of 1 *www-wds.worldbank.org.*
2. For more information on the work and mandate of the WCD, see *www.dams.org.*
3. According to the International Commission on Large Dams, a large dam is 15 metres or higher. Dams between 5 and 15 metres with a reservoir volume of more than 3 million cubic metres are also classified as large dams.
4. The findings of the commission were published in the report "Dams and development: A new framework for decision making". The report can be downloaded from *www.dams.org.*
5. Nilsson C, C A Reidy, M Dynesius, & C Revenga (2005), Fragmentation and flow regulation of the world's large river systems, Science 308:405-408.
6. WWAP (World Water Assessment Programme), 2003, Water for People Water for Life, The United Nations World Water Development Report, United Nations Educational, Scientific & Cultural Organisation and Berghahn Books, Barcelona.

7. Millennium Ecosystem Assessment, 2005, Ecosystems and Human Well-being: Synthesis. Island Press, Washington, DC.
8. Syvitski J M P, Vorosmarty C J, Kettner A J, Green, P (2005), Impact of Humans on the Flux of Terrestrial Sediment to the Global Coastal Ocean, Science, Vol. 308, 15 April 2005, pp. 376 to 380.
9. *International Journal on Hydropower & Dams* (2005), World atlas & industry guide.
10. Minty (2001), A wildlife impact assessment for the proposed Macal River upper storage facility *www.probeinternational.org.*
11. Conservation Strategy Fund (2000), Analysis of the Final Feasibility Study and Environmental Impact, Assessment for the Proposed Chalillo Dam, *www.conservation-strategy.org.*
12. Privy Council Appeal No. 47 of 2003, *www.stopfortis.org.*
13. International coalition to save the Macal River, Stop Fortis Campaign -Update August 2005, *www.stopfortis.org.*
14. Matola S (2005), Belize Zoo, Personal communication.
15. Hershowitz A (2005), Natural Resources Defense Council, Personal communication.
16. See Belize Public Utilities Commission (2005) *www.puc.bz www.probeinternational.org.*
17. WWF Turkey (2003), Ermenek Dam & Hydropower Project, Evaluation of ecological issues.
18. The Bern Convention is a binding international legal instrument in the field of nature conservation, which covers the whole of the natural heritage of the European continent and extends to some States of Africa. Its aims are to conserve wild flora and fauna and their natural habitats and to promote European cooperation in that field.
19. OECD (2005), Economic Survey of Iceland, 2005. *www.oecd.org.*
20. Solnes J (2003), Environmental quality indexing of large industrial development alternatives using AHP, in: Environmental Impact Assessment Review, 23/2003, pp. 283-303.
21. The project is expected to generate annual revenues of US$13 million in the first year of operation with annual revenue growing to about US$150 million by the year 2033. *web.worldbank.org.*
22. See for example Graecen & Sukkamoed (2005), Did the World Bank fudge figures to justify NT2, *www.palangthai.org.*

    Probe International (2004) PI News Release: World Bank report confirms Nam Theun 2 is not Thailand's least-cost power option, *www.probeinternational.org.*
23. World Bank (2005), The Nam Theun 2 Hydroelectric project. An overview and update, April 2005, *siteresources.worldbank.org.*

24. *www.dams.org.*

25. Confederación Hidrográfica del Guadalquivir (1995) Evaluación de Impacto Ambiental del Embalse de los Melonares, Sevilla.

26. *www.panda.org.*

27. WFD Provisional Art.5 Report of the Guadalquivir RBA River Basin Authority, *www.chguadalquivir.es.*

28. WWF Australia (2004), WWF Submission – Assessment issues in the 2004 NCP Assessment Framework for Water Reform.

29. Armstrong (2004), Where wild things are dammed. In: ECOS Issue 122, *wnloads.publish.csiro.au.*

30. Further details on the Dams and Development Project can be found on *www.unep.org.*

31. IHA (2004), Sustainability Guidelines *www.hydropower.org.*

32. Hildyard N and Gilfenbaum E (2005), The OECD Arrangement and New Subsidies for Dams: The Case for Strengthened Standards. *www.thecornerhouse.org.uk.*

33. See *www.hsbc.com.*

34. WWF Brazil (2003), Repowering hydroelectric utility plants as an environmentally sustainable alternative to increasing energy supply in Brazil.

6

Managing the Environmental Impact of Dams

Matthew P McCartney and Hilmy Sally

In its final report, published in November 2000, the World Commission on Dams concluded that the benefits derived from large dams have made a significant contribution to human development. However, in many cases, the environmental and social costs have been unnecessary and, by present standards, unacceptable. Often negative impacts arise as a consequence of lack of foresight and because dams are planned and managed in isolation from other developments occurring in a catchment area. Given the huge number of existing dams and the large number that may be built in the future, it is clear that humankind must live with the environmental and social impacts for many decades to come. Consequently, there is a need to improve environmental practices in the operation of both existing and new dams. This paper provides a brief review of the consequences for ecosystems and biodiversity resulting directly from the presence of dams on rivers. Strategies to protect the environment are described. A prerequisite for successful environmental protection is that dams are managed within the specific environmental, social and economic context of the catchment areas in which they are situated.

Introduction

Dams represent one of the most significant human interventions in the hydrological cycle. Through provision of water for drinking, irrigation and electricity, they have supported human socioeconomic development, but simultaneously they have had a considerable impact on freshwater ecosystems. Where water is over-extracted, its quality degraded or hydrological regimes modified, the natural environment deteriorated, habitats are destroyed and ecological functions, many of which enhance peoples well-being, are lost.

It is estimated that inter-basin transfers and water withdrawals for supply and irrigation have fragmented 60% of the world's rivers (Revenga *et al.*, 2000). For most of the world's existing stock of dams, environmental issues played little part in their design and operation. However, in the last two decades, an increase in environmental awareness has led to the recognition that the management of water resources includes a responsibility to protect the users of water, and the natural resources that depend on water, from over-utilisation or impacts that cause degradation. As a result, considerable effort has been invested in developing approaches to lessen the most damaging effects of dams. However, experience indicates that the success of these measures is extremely variable and far from assured (Bergkamp *et al.*, 2000).

Abiotic Impacts of Dams

Rivers exist as a continuum of linked surface and groundwater flow paths and are important natural corridors for the flows of energy, matter and species. The spatial and temporal heterogeneity of river systems is responsible for a diverse array of dynamic aquatic habitats and hence biological diversity, all of which are maintained by the constantly changing flow regime. Inundation of flood plains increases organic matter decomposition and nutrient cycling and has led to the evolution of adaptive strategies that are tightly coupled to the flood regime.

Dams constitute obstacles for longitudinal exchanges along fluvial systems and so result in "discontinuities" in the river continuum (Ward and Stanford 1995). Post-impoundment phenomena directly and indirectly influence a myriad of factors that affect natural processes and so, ultimately, alter the ecological structure of ecosystems, sometimes tens or even hundreds of kilometres downstream.

Impacts on Flow Regime

The most obvious impact of storage reservoirs is the upstream inundation of terrestrial ecosystems and, in the river channel, the conversion of lotic to lentic systems. Dams also alter the downstream flow regime. The effect of a dam and its reservoir on flow regimes depends on both the storage capacity of the reservoir relative to the volume of river flow and the way the dam is operated. The most common attribute of flow regulation is a decrease in the magnitude of flood peaks and an increase in low flows. A consequence of reduced flood peaks is reduction in the frequency and extent of overbank flooding. For example, in the Hadejia-Nguru wetlands in Nigeria, annual flooding of about 3,000 km^2, prior to the building of dams was reduced to less than 1000 km^2 after construction (Hollis *et al.*, 1993). In some circumstances, operational procedures can result in rapid flow fluctuations that occur at non-natural rates. Hydroelectric power and irrigation demands are the most usual causes, but short-duration high discharges are also utilised for navigational purposes and for recreation. For many purposes, the so-called "pulse releases" are made regularly. For example, daily releases through power turbines often reflect diurnal variation in power demand.

Impacts on Thermal Regime

Water temperature influences many important ecological processes. Temperature is an important factor affecting growth in freshwater fish, both directly and indirectly, through feeding behaviour, food assimilation, and the production of food organisms. Under natural conditions the relatively small volume of water in a river section and turbulent mixing ensure that river water responds rapidly to changes in the prevailing meteorological conditions. In contrast, the relatively large mass of still water in reservoirs allows heat storage and produces a characteristic seasonal pattern of thermal behaviour. Depending on geographical location, water retained in deep reservoirs may become stratified. Releases of cold water from the hypolimnion (i.e., the deep cold layer) of a reservoir, is the greatest non-natural consequence of stratification. However, even without thermal stratification, water released from reservoirs is often thermally out of phase with the natural regime of the river.

Impacts on Chemistry

Water storage in reservoirs induces physical, chemical and biological changes, all of which affect water chemistry. Consequently the water discharged often has a very different composition to that of inflowing rivers. Nutrients, particularly phosphorous, are released biologically and leached from flooded vegetation and soil. Oxygen demand and nutrient levels generally decrease as the organic matter decays.

Some reservoirs require many years for the development of stable water-quality regimes. After maturation reservoirs can, like natural lakes, act as nutrient sinks. For example, in comparison to the inflows, mean concentrations of orthophosphate in the outflows from the Callahan Reservoir, Missouri, USA, were reduced by 50% (Schreiber and Rausch 1979). Eutrophication of reservoirs may occur as a consequence of large influxes of organic material and nutrients, often arising as a consequence of anthropogenic activity in the catchment (Chapman 1996). Hence catchment management has a key role to play in sustaining reservoir water quality.

The quality of water released from a reservoir is determined by the elevation of the outflow structure(s). Water released from near the surface is generally well-oxygenated, warm, nutrient-depleted water. In contrast, water released from near the bottom is often cold, oxygen-depleted, nutrient-rich water that may be high in hydrogen sulphide, iron and manganese.

Bacterial decomposition of material in reservoirs can transform inorganic mercury into methylmercury, a toxin of the central nervous system. Bioaccumulation results in levels of methylmercury in the tissues of fish at the top of the food-chain several times higher than in small organisms at the bottom of the food-chain (Bodaly *et al.,* 1984). This can have serious implications for people that depend on fish for a large proportion of their diet. For example, mercury levels in hair samples of Cree Indians in the James Bay region of Quebec in Canada, were found to be above the World Health Organizations-recommended upper limit (i.e., 6 ppm by weight) as a consequence of eating fish from reservoirs (Dumont 1995).

Impacts on Sedimentation

Reservoirs reduce flow velocity and so enhance sedimentation. The rate at which sedimentation occurs within a reservoir depends on the physiographic features and land-use practices of the catchment, as well as the way the dam is operated. Large magnitude and frequent fluctuation in water levels in reservoirs can cause erosion of the shores and add to deposition. It is estimated that between 0.5% and 1% of the storage volume of the world's reservoirs is lost annually due to sediment deposition (Mahmood 1987).

Downstream of a dam, reduction in sediment load in rivers can result in increased erosion of river-banks and beds, loss of floodplains (through erosion and decreased over-bank accretion) and degradation of coastal deltas. Removal of fine material may leave coarser sediments that armour the riverbed, protecting it from further scour. In some circumstances, material entrained from tributaries cannot be moved through the channel system by regulated flows, resulting in aggradation. Reservoir flushing (i.e., the selective release of highly turbid waters) is a technique sometimes used to reduce in-reservoir sedimentation. Consequently, reservoir operations may periodically result in unnaturally high concentrations of sediment in downstream systems.

Impacts on Organisms and Biodiversity

Dams, through disruption of physiochemical and biological processes, modify the conditions to which ecosystems have adapted. The impacts of dams vary substantially from one geographical location to another and are dependent on the exact design and the way a dam is operated. Every dam has unique characteristics and, consequently, the scale and nature of environmental changes are highly site-specific. However, impacts invariably affect biota and can impact biodiversity.

Impacts on Primary Production

The introduction of a dam into a river system affects primary production. In freshwater ecosystems, phytoplankton, periphyton and macrophytes form the base of the foodweb. Upstream of a dam, the slow-moving water of the reservoir is often an ideal habitat for phytoplankton, but, depending on depth, temperature, light penetration and the nature of the substrate, may be less suited for periphyton and rooted macrophytes. Downstream of a dam, primary production is affected by the changes to flow, water chemistry and thermal regimes, as well as current velocities and turbidity. In many temperate climates, increased summer flows, higher water temperatures in winter, reduction of turbidity, decreased scouring of the substrate and reduced effluent dilution often enhance primary production. Modification of primary production may alter the aquatic environment directly. For example, blooms of phytoplankton and floating plants (e.g., water hyacinth) reduce light penetration and deplete oxygen when they decompose, and so have an adverse impact on other species (Joffe and Cooke 1997).

Dams can also affect riverside and floodplain vegetation, the characteristics of which are often controlled by the dynamic interaction of flooding and sedimentation. By changing the magnitude and extent of floodplain inundation and land-water interaction, dams can disrupt plant reproduction and allow the encroachment of upland plants previously prevented by frequent flooding. Studies in Norway have shown that the presence of storage reservoirs permanently reduces the diversity of riparian vegetation (Nilsson *et al.*, 1997).

Impacts on Fish

Few fish are adapted to both lotic and lentic habitats. Consequently, the transformation of a river to a reservoir often results in the extirpation of resident riverine species. Downstream of dams, marked changes in fish populations occur as a consequence of blockage of migration routes, disconnection of the river and floodplain and changes in flow regime, physiochemical conditions (e.g., temperature, turbidity and dissolved oxygen), primary production and channel morphology. These changes may benefit some species but they generally have an adverse effect on the majority of native species.

The 1996 IUCN Red List of Threatened Animals includes 617 freshwater fishes (i.e., about 6% of the known number of freshwater species). Other researchers have speculated that globally between 20% and 35% of all freshwater fish are threatened (Staissny 1996). Although the loss of species is not solely a consequence of dams, they are one of the principal factors. It is estimated that half the fish stocks endemic to the Pacific coast of the USA have been lost in the past century to a large extent because of dam construction (Chaterjee 1998).

Impacts on Birds and Mammals

The importance of riparian corridors for birds and terrestrial animals has been demonstrated (e.g., Decamps *et al.,* 1987). The creation of reservoirs has both positive and negative effects for aquatic and terrestrial species. The inundation of ecosystems inevitably leads to the loss of habitat and terrestrial wildlife. In tropical areas, flooding forests high in endemic species extirpates many and, in some circumstances, may result in species extinction. In contrast, in arid climates, reservoirs provide a permanent water resource that may benefit many species. In South Africa, the presence of reservoirs has greatly increased the availability of permanent water bodies, and has had a major effect on the distribution and numbers of waterfowl (Cowan and Van Reit 1998).

The most negative downstream consequence of river regulation on mammals and birds is the disruption of the seasonal flood regime along the river (Nilsson and Dynesius 1994). In the long-term, reduced flooding can alter vegetation communities that may be important for a wide range of mammal and bird species. In arid regions, riparian vegetation may be the only significant vegetation, and many animals will have adapted behavioural patterns to fit with seasonal flooding. If the flooding regime is altered, changes in vegetation may place at risk the birds and animals that depend on it.

Environmental Protection

Options for Environmental Protection

Engineers, environmental scientists and ecologists have developed a broad range of technical and socioeconomic interventions to ameliorate the most damaging impacts of dams. For new dams, these can be conceptualised within a hierarchical framework comprising three types of measure:

- Avoidance measures result in no change to the existing environmental functioning of a particular area by avoiding anticipated adverse effects. For dams this means alternatives to dam construction such as demand management, water recycling, rainfall harvesting or alternatives to hydropower (e.g., solar, wind, thermal or nuclear). All alternatives have economic, social and environmental consequences that must be weighed against those arising from dam construction.

- Mitigation measures reduce the undesirable effects of a dam by modification of its structure or operation, or through changes to the management of the catchment within which the dam is situated. To date, mitigation is the most widely used approach to ameliorating the negative impacts of dams and a wide range of technical interventions has been developed (Tables 1 and 2). For example, making environmental flow releases to sustain downstream ecosystems is increasingly common. However, to be successful in a specific situation, mitigation measures require a great deal of understanding of complex processes and their interactions. Strategies are often of limited effectiveness, or may even result in undesirable effects, if detailed scientific and engineering studies are not conducted beforehand.
- Compensation measures compensate for effects that can neither be avoided nor sufficiently mitigated. Principal approaches include preservation of existing ecologically important areas (e.g., through the establishment of a national park) and rehabilitation of previously disturbed land either around reservoirs or some distance from the development in question.

Ideally, environmental protection measures are identified through an Environmental Impact Assessment, so that, adverse affects are minimised from the outset of a project. In many situations integrated approaches that incorporate changes in catchment management are essential for success.

Constraints to Successful Environmental Protection

Measures to protect the environment are successful in some circumstances but are not effective in others (Bergkamp *et al.*, 2000; IEA 2000). Constraints to successful environmental protection are not limited to technical deficiencies but also arise because of limitations in human, financial and institutional capacity.

At present, lack of scientific understanding is one of the primary constraints to successful environmental protection. Notwithstanding the research conducted to date, it is often impossible to predict, even with site-specific studies, what many of the precise impacts of a dam will be. There is still very little knowledge of the habitat requirements of many species. The relationships between biophysical and socioeconomic aspects of systems are even less well-understood and so often the social implications of the alteration of ecosystems cannot be foreseen.

Developing the scientific and socioeconomic knowledge base required to successfully ameliorate impacts, requires comprehensive field investigations, necessitating significant time and financial resources. In many projects, funds for conducting environmental impact assessments and for post-project monitoring are insufficient.

The responsibilities for planning, monitoring and regulation of dams are often spread across a large number of institutions. Disparate organisation complicates management co-ordination and the identification of responsibility. This is a problem that is exacerbated in those countries where there is neither the necessary framework to ensure legal compliance, nor a civil society sufficiently empowered to insist that recommended measures to protect the environment are put into practice.

Table 1: Measures Upstream and within the Reservoir to Mitigate the Impact of Dams on Ecosystems

Issue	Mitigation Measure	Examples
Thermal regime	Changes to inlet structure configuration. Artificial mixing by mechanical mixer or a compressed air. Flushing to reduce residence times.	Automatic aeration, controlled by temperature sensors was installed at the Teddington dam in Australia in 1996. This maintains unstratified well oxygenated conditions and prevents high manganese concentrations in the raw water supply (Burns, 1998).
Water quality	Catchment management. Pre-impoundment clearing of reservoir. Reservoir re-aeration. Treatment of reservoir inflows. Flushing to reduce residence times. Construction of small "pre-reservoirs".	To reduce eutrophication in the Cirata and Saguling reservoirs in Indonesia, a program of urban and industrial wastewater treatment within the upstream catchment has been proposed (Simeoni *et al.*, 2000). Five pre-reservoirs (i.e., small reservoirs with a retention time of a few days) have been constructed upstream of the main Eibenstock reservoir in Germany to improve water quality and reduce sedimentation in the main reservoir (Putz and Bendorf reservoirs, 1998). At Grafham Water in the UK, influent water was dosed with ferric sulphate to reduce reservoir phosphorous concentrations and so reduce algal concentrations (Daldorph, 1998).

Contd...

Contd...		
Sedimentation	Catchment management. Debris dams. Shoreline erosion control. Sediment flushing. Utilisation of sediment density currents. Dredging.	At the Fortuna dam in Panama, a 10 km² reservoir is surrounded by a 160 km² natural reserve. This limits erosion and reduces sediment deposition in the reservoir (Leibenthal 1997). Oulujarvi, a regulated lake in Finland, is drawn down to reduce the erosion of sandy shores caused by spring floods (Hellesten 1996). Sediment flushing of the Hengshan reservoir in China, for a few weeks every 2-3 years, enables the long-term capacity of the reservoir to be maintained at 75% of the original capacity (Atkinson 1996).
Weeds	Mechanical cutting. Chemical control. Biomanipulation.	An integrated management strategy has been developed to control water hyacinth in the Yacyreta reservoir on the Parana River in Argentina. This includes biomass clearing, development of effective sewage treatment plants to reduce nutrient input to the reservoir and a program of water releases (Joffe and Cooke 1997).
Fish	Man-made spawning areas. Removal of sand bars across tributary mouths. Construction of shallow habitat. Introduction of lake species into reservoir.	New spawning grounds were successfully created in the upgrading of the Riviere-des-waterPrairies project in Canada (IEA 2000). More than 1.5 million fish (i.e., salmon, rainbow trout and brook trout) were introduced into the Williston Reservoir in British Columbia, Canada (IEA 2000).
Terrestrial wildlife	Wildlife rescue. Enhancement of reservoir islands for conservation.	10,000 animals were rescued from drowning prior to the filling of the Afokaba reservoir on the Surinam River in South America (Nilsson and Dynesius 1994).

Table 2: Measures to Mitigate the Downstream Impact of Dams on Ecosystems

Issue	Mitigation measure	Example
Flow regime	Managed flow releases.	The Physical Habitat Simulation System (PHABSIM) has been used to compare options for minimising the in-stream ecological impacts of river regulation through compensation flow releases from the Derwent Valley Reservoir System in the UK (Maddock *et al.*, 2001).
Thermal regime	Multi-level outlet works.	A multiple level outlet tower has been proposed for the Glen Canyon dam in the USA to mitigate the impact of cold water releases on trout (CGER 1996).
Water quality	Outlet works aeration. Multi-level outlet works. Turbine venting.	In the USA Duke Power has experimented with various approaches to increasing dissolved oxygen levels in turbine tailraces. At the Wateree Dam, turbine blades were modified to enable air to be drawn into the water through small holes in the turbine vanes. This produced a 3 mgl^{-1} increase in DO, without significantly impacting turbine performance (Sigmon *et al.*, 2000).
Sedimentation	Addition of sediment to rivers. Managed flow releases. Shoreline stabilisation. Pumping offshore sediment to estuaries.	Gravel has been added to the River Rhine, since 1977 downstream of dams, to reduce erosion and maintain the channel morphology (Dister *et al.*,1990). On the Galuare River in France banks historically protected with rip-rap, are now being protected through the regeneration of a buffer zone of riparian woodland (Piegay *et al.*, 1997).
Weeds/algal blooms	Mechanical cutting. Chemical control. Biological control. Flushing.	Mechanical harvesting of aquatic macrophytes was attempted on the River Otra in southern Norway, to control *Juncus bulbosus*. The approach was largely ineffective because of high perational costs and inadequate removal of submergent vegetation (Rørslett and Johansen 1996). Research has shown that algal blooms on the Murray River, Australia can be dispersed through a combination of flow management and reduction in water-levels behind weirs (Maier *et al.*, 2001).

Contd...

Contd...		
Fish	Freshets to stimulate fish migration. Improved design of turbine, spillways and overflows. Fish passes. Artificial spawning areas.	A vertical slot fish pass has been shown to be effective in enabling 24 species of fish, including barramundi *(Lates calcarifer)* to move upstream of the Fitzroy barrage in Australia (Stuart and Mallen-Cooper 1999).
	Hatcheries and fish stocking.	The hydropower dam in the Hunderfossen project in Norway was a barrier to migratory trout. A fish ladder was unsuccessful, because the primary constraint to fish migration was reduced downstream flows. Trout restocking also proved to be less successful than expected. An increase in minimum downstream flows at certain times of year to trigger migration has improved the situation (IEA 2000).
Terrestrial wildlife	Managed flow releases.	High flow releases were designed into the operation of the Itezhi-tezhi dam in Zambia. One reason for this was to preserve, through annual flooding, the high biodiversity of the internationally important Kafue Flats (McCartney 2002).

Compliance with Commitments for Environmental Protection

A broad body of regulation and guidelines applicable to the environmental impacts of large dams exists at national and international levels. In addition to those developed by the International Commission On Large Dams (ICOLD) and the International Energy Agency (IEA), most multilateral and bilateral development financing agencies now have comprehensive policies that cover environmental issues. However, incorporating environmental protection measures into large dam projects is made difficult by the failure of many developers and operators to fulfill voluntary and mandatory obligations. Principal causes for this have been identified (WCD 2000) as:

- Lack of, and incompleteness in, policy, legal and regulatory frameworks.
- Difficulties in accurately defining environmental requirements and specifying these in the implementation agreements of projects.

- Lack of human, financial and organisational capacity for project appraisal and to act on infringements of agreements.
- Lack of transparency and accountability.
- Weak or non-existent recourse and appeals mechanisms.

Improving compliance requires incentives and sanctions as well as mechanisms for monitoring environmental performance. Furthermore, there is need for greater consistency in the criteria and standards stipulated by different funding agencies as well as increased transparency and accountability in the decision-making process. To deal with these issues, it has been proposed that the dam industry should adopt an ethical code of conduct to ensure that environmental concerns are adequately addressed and human rights respected. Such a code would provide guidance for environmental management, public participation and conflict resolution at each stage of project development and operation (Lafitte 2001).

Regular environmental auditing, by independent bodies, leading to certification, has been proposed for both existing and new dams (WCD 2000). Environmental management is a prerequisite for certification and the development of an ISO (International Organisation for Standardisation) standard for dam management is being contemplated (Giesecke *et al.*, 2000). Compliance plans that specify binding arrangements for specific social and environmental commitments are one way of encouraging developers and operators to implement environmental protection measures (WCD 2000). In North America licensing of dams is an important mechanism for initiating environmental protection measures. Relicensing (typically every 25 to 30 years) is now often made conditional on improved environmental protection, reflecting contemporary priorities.

Conclusion

The management of natural resources and particularly freshwaters will be a key human endeavour in the 21st century. Given the large number of existing dams and those that may be built in the future, it is clear that humankind must live with the environmental and social consequences for many decades to come. Most dams are built with the best of intentions: to provide water supplies and power at times when water is naturally scarce and to reduce the devastating effects of floods. These are all worthy reasons for river regulation. However, it is now recognised that if development is to be sustainable, the effects of impoundment on ecosystems and other species cannot be neglected. Minimising the negative environmental effects of dams must become a prime focus of attention by owners, operators, financial institutions and environmental managers.

A prerequisite for sustainable development is that future dam planning, construction and operation must become part of an integrated management effort that gives prominence to environmental protection. All the environmental impacts of a dam should be evaluated within the specific environmental, social and economic context of the catchment in which it is located. This requires inter-disciplinary thinking and basic understanding of the complex interactions between ecological and socioeconomic systems. This is particularly true of environmental flow releases, where lack of hydro-ecological understanding remains a key constraint to successful implementation. There is an urgent need for further research to link abiotic processes and the impact of dams on these processes to ecological change and the socioeconomic consequences.

(Dr. Matthew P McCartney is Senior Researcher (Hydrologist) at IWMI East Africa and Nile Basin (Ethiopia).

Dr. Hilmy Sally is Senior Researcher and head of the Regional Office of the International Water Management Institute (IWMI), Pretoria, South Africa. He has 25 years of working experience in water resources and, irrigation research and development in various countries of Africa, Asia and Europe.)

References

Atkinson E 1996, The feasibility of flushing sediment from reservoirs, Hydraulics Research, Wallingford.

Bergkamp G, McCartney M, Dugan P, McNeely J, Acreman M 2000, Dams, ecosystem functions and environmental restoration: Thematic Review II.1 World Commission on Dams, Cape Town.

Bodaly R A, Hecky R E, Fudge R J P, 1984, Increases in fish mercury levels in lakes flooded by the Churchill River diversion, northern Manitoba, *Canadian Journal of Fisheries and Aquatic Sciences* 41, 682-691.

Burns F L, 1998, "Case study: automatic reservoir aeration to control manganese in raw water. Maryborough town water supply Queensland, Australia", *Water Science and Technology* 37, 301-308.

Chapman M A, 1996, Human impacts on the Waikato River System, New Zealand, *Geojournal,* 40, 85-99.

Chaterjee P 1998, Dam busting, *New Scientist* 34-37.

Commission on Geosciences, Environment and Resources (CGER), 1996. River Resource Management in the Grand Canyon. Report of the Committee to Review the Glen Canyon Environmental Studies.

Cowan G A, van Riet W 1998, A directory of South African wetlands, Department of Environmental Affairs and Tourism, Pretoria.

Daldorph P W G 1998, Management and treatment of algae in lowland reservoirs in Eastern England, Water Science and Technology 37, 57-63.

Decamps H Joachim J, Lauga J 1987, The importance for birds of the riparian woodlands within the alluvial corridor of the river Garonne, S W France, Regulated Rivers: Research and Management 1, 301-316.

Dister E, Gomer D, Obrdlik P, Petermann P, Schneider E 1990, Water management and ecological perspectives of the Upper Rhine's floodplains, Regulated Rivers: Research and Management 5, 1-15.

Dumont C 1995, Mercury and health: The James Bay Cree experience, Proceedings of the Canadian Mercury Network Workshop, 1995 (Montreal: Cree Board of Health and Social Services).

Giesecke J, Heimerl S, Markard J, Keifer B 2000, Environmental certification systems for hydropower: ISO 14001, EMAS and third party ecolabelling schemes. In Hydro 2000: Conference Proceedings, *International Journal of Hydropower and Dams* 727-736.

Hellesten S, Marttunen M, Palomaki R, Riihimaki, J Alasaarela, E 1996, Towards an ecologically based regulation practice in Finnish Lakes. Regulated Rivers: Research and Management 12,: 535-541.

Hollis G E, Adams W M, Kano M A 1993, The Hadejia-Nguru wetlands: Environment, economy and sustainable development of a Sahelian floodplain wetland, IUCN, Gland.

International Commission on Large Dams (ICOLD) 1997, Position paper on dams and the environment. *http://genepi.louis-jean.com/cigb/chartean.html*

International Energy Agency (IEA) 2000, Survey of the environmental and social impacts and the effectiveness of mitigation measures in hydropower development, IEA 92 pp.

Joffe S, Cooke S 1997, Management of the water hyacinth and other invasive aquatic weeds: issues for the World Bank, The World Bank, Washington DC.

Lafitte R 2001 Ethics and dam engineers, I*nternational Journal of Hydropower and Dams* 4, 58-59.

Leibenthal A 1997, The World Banks experience with large dams: a preliminary review of impacts, The World Bank, Washington DC.

Maddock I P, Bickerton M A, Spence R Pickering, T 2001. Reallocation of compensation releases to restore river flows and improve instream habitat availability in the Upper Derwent catchment, Derbyshire, UK, Regulated Rivers: Research and Management 17, 417-441.

Mahmood K 1987, Reservoir Sedimentation impact, extent and mitigation, World Bank Technical Paper No. 71, World Bank, Washington DC.

Maier H R, Burch M D, Bormans M, 2001, Flow management strategies to control blooms of the cyanobacterium, *Anabaena circinalis*, in the River Murray at Morgan, South Australia. Regulated Rivers: Research and Management 17, 637-650.

McCartney M P 2002, Large dams and integrated water resources management, with reference to the Kafue hydroelectric scheme. In Hydro 2002: Conference Proceedings, *International Journal of Hydropower and Dams* 389-397.

Nilsson C, Jansson R, Zinko U 1997, Long-term responses of river-margin vegetation to water-level regulation, Science, 276, 798-800.

Nilsson C, Dynesius M, 1994, Ecological effects of river regulation on mammals and birds: a review, Regulated Rivers: Research and Management, 9, 45-53.

Piegay H, Cuaz M, Javelle E, Mandier P 1997, Bank erosion management based on geomorphoogical, ecological and economic criteria on the Galuaure River, France. Regulated Rivers: Research and Management 13, 433-448.

Putz K, Benndorf J 1998, "Importance of pre-reservoirs for the control of eutrophication of reservoirs", *Water Science and Technology* 37, 317-324.

Revenga C, Brunner J, Henninger N, Kassem K, Payne R, 2000, Pilot analysis of global ecosystems: Freshwater Systems, World Resources Institute, Washington DC.

Rørslett B, Johansen S W 1996, Remedial measures connected with aquatic macrophytes in Norwegian regulated rivers and reservoirs, Regulated Rivers: Research and Management 12, 509-522.

Schreiber J D, Rausch D L 1979, Suspended sediment phosphorous relationships for the inflow and outflow of a flood detention reservoir, *Journal of Environmental Quality* 8, 510-514.

Sigmon J C, Lewis G D, Snyder G A, Beyer J R 2000, Improving water quality by application of turbine aeration a case study. In Hydro 2000: Conference Proceedings, *International Journal of Hydropower and Dams* 417-425.

Simeoni G, Hanselmaan K, Harnanto M, Semiawan A 2000, Trends of pollutant indicators and mass balances in tropical hydropower reservoirs. In: Hydro 2000: Conference Proceedings, *International Journal of Hydropower and Dams* 2000, 409-416.

Staissny M L J 1996, An overview of freshwater biodiversity: with some lessons from African fishes, *Fisheries* 21, 7-13.

Stuart I G, Mallen-Cooper M 1999, An assessment of the effectiveness of a vertical-slot fishway for non-salmonid fish at a tidal barrier on a large tropical/subtropical river, Regulated Rivers: Research and Management 15, 575-590.

Ward J V, Stanford J A 1995, Ecological connectivity in alluvial river ecosystems and its disruption by flow regulation. Regulated Rivers: Research and Management 11, 105-119.

World Commission on Dams (WCD) 2000, Dams and Development: a new framework for decision-making, Earthscan Publications Ltd., London.

7

Narmada and the Myth of Rehabilitation

Mike Levien

The report of the Shunglu Committee appointed to verify rehabilitation measures accorded to families affected by the Sardar Sarovar Project is riddled with several contradictions and glaring inadequacies. Though there is clear recognition of the incomplete process of rehabilitation, the report chooses to ignore the Supreme Court rulings on the issue as well as the assurances made by several state governments. Moreover, the fact that the government chose to set up the committee even while it allowed construction to go ahead gives the lie to all its pious declarations of ensuring a judicious rehabilitation. On a wider level, the report symbolises once again the connections between "power" and "truth"; it is the project-affected families that have been rendered marginal, in shaping either the truth or the country's economic development.

As the monsoon rains begin to fall, labourers can be seen working on bleak concrete structures on a dusty patch of land in the Nimad plains of Madhya Pradesh. The scene is striking in two respects. First, is the profound contrast between these desolate and barren "rehabilitation sites" for the Sardar

Source: Economic and Political Weekly, August 19, 2006. (www.epw.org.in/). © EPW. Reprinted with permission.

Sarovar Project (SSP) and the fertile and vibrant villages that dam-affected families are being forced to leave. Second, is the amazing fact that construction of the sites is still far from complete. Even rudimentary houseplots at many sites are unfinished, not to mention civic amenities and, most importantly, the provision of irrigable, cultivable land to displaced families. The law, as laid down by the Narmada Water Disputes Tribunal Award (NWDTA) of 1979, and previous Supreme Court decisions, very clearly mandate that the dam construction cannot proceed unless all of these are provided one year *before submergence*. Yet, with the rains falling and the large-scale submergence of up to 35,000 families imminent, even the façade of rehabilitation is incomplete. Both the Supreme Court and prime minister Manmohan Singh have chosen to ignore the patent illegality and injustice of this situation and have allowed dam construction to continue up to 122 metres. As a result, thousands of families in the Narmada Valley will face submergence this August with nowhere else to turn. They are among the millions falling victim to the lethal combination of a flawed development paradigm and a government that refuses to respect the legal and human rights of its citizens.

Satyagraha and the Government's Betrayal

Sardar Sarovar returned to claim national attention again last April with a month-long dharna by the Narmada Bachao Andolan (NBA) in Delhi, including a 20-day fast by Medha Patkar, Jamsingh Nargave and Bhagwatibai Patidar. Forcibly hospitalised, famished, Patkar and the NBA resolutely demanded that the government halt dam construction until rehabilitation is completed as required by law. The escalation of tactics was a reflection of the tremendous stakes that hung on the decision of whether to raise the dam height, affecting the homes and livelihoods of approximately 35,000 families who would be in the submergence zone at the 122m dam height. Support poured in from all quarters, and solidarity protests were held in cities across India and embassies across the world. Coming after 20 years of persistent agitation and struggle, innumerable fasts, protests, and monsoon satyagrahas, the dharna was a tremendous display of courage and persistence, and a demonstration of the incredibly strong solidarity the NBA has inspired people across the country.

Nonetheless, in the face of overwhelming evidence that rehabilitation in the Narmada Valley is incomplete, the Supreme Court and the prime minister failed in their responsibility to uphold the law and stop dam construction. In Madhya

Pradesh's own submission to the court, it admitted that construction on 11 out of 86 rehabilitation sites was still incomplete (itself an underestimation). The group of ministers (GoM) team, appointed by the prime minister to investigate rehabilitation in the valley, found upon their visit that rehabilitation was far from accomplished. The ministers, lead by water resource minister Saiffudin Soz, released a report on April 17 in which they noted "extreme shortcomings" in the resettlement and rehabilitation of oustees and estimated that it would take at least another year for rehabilitation to be completed if the government showed the political will to accomplish this "gigantic task." In the NBA's own brief to the court, it thoroughly documented the total absence of land-based rehabilitation in Madhya Pradesh, the exclusion of thousands of people from the official Project Affected Families (PAF) lists, and the faulty and incomplete nature of even the 75 rehabilitation sites that the government claims to have constructed. It is beyond dispute that rehabilitation in the Narmada Valley is incomplete.

It is also indisputable that rehabilitation must be completed before construction on the dam can legally continue. According to clause XI of the NWDTA, rehabilitation of all PAFs in all three affected states (Madhya Pradesh, Gujarat, and Maharashtra) must be provided for one year before each successive increase in the dam height. The Supreme Court has itself upheld this position in its decisions of 2000 and 2005. The Court's 2005 decision clearly states, "In terms of NWDT awards the irrigable lands and house sites were required to be made available to PAFs one year in advance of submergence, and requisite amenities were also to be provided. Further, the notices for vacation of the lands are to be given after completion of R&R of PAFs on or before December 31, i.e., six months before actual submergence, (likely on July 1 of the next year). In terms of these stipulations, raising of the dam, which would cause submergence would not be permitted unless rehabilitation programme is carried out." The judgment thus clearly and unambiguously states that no construction can be allowed on the dam unless rehabilitation sites are completed one year before submergence and PAFs shifted no later than six months before submergence. Yet, the court has inexplicably failed to uphold even its own judgment.

In its 2005 decision, the court also upheld the principle laid down in the NWDTA that rehabilitation must be land-based. Cash compensation, which the Madhya Pradesh government (GoMP) has been distributing in lieu of land, is explicitly prohibited by the NWDTA, which mandates that all PAFs be given two hectares of cultivable and irrigable land either in the command area of the dam or in their own state (according to the oustees choice). Land-based rehabilitation is also mandated by ILO Convention 107 on indigenous people, to which the Indian government is a signatory. The convention clearly states that in the exceptional circumstances in which indigenous people are to be forcibly removed from their lands, they must be given replacement land that is as good as, if it not better than, that from which they are removed. Thus, cash compensation, as per the NWDTA, the Supreme Court's own decision, and the ILO, is completely illegal for the SSP. Nonetheless, GoMP refuses to recognise this fact and continues to illegally distribute cash instead of land. Amazingly, in Madhya Pradesh, where 193 of the 245 villages affected by the SSP are located, almost no PAFs have been given replacement cultivable land!

Thus, the petition filed in the Supreme Court by 48 dam-affected villagers from Madhya Pradesh was air-tight. The law is indisputable that rehabilitation must be totally completed and all villagers totally shifted six months before submergence. The facts were clear that rehabilitation was incomplete, even by GoMP's own admission, not to mention the much graver shortcomings identified by the GoM and the affected people. Thus, both the law and the facts were clear, with the one standing in indisputable contradiction to the other. Unfortunately, this is not always enough. The highest court in the land refused to enforce its own previous decisions and enforce the law. The PM refused to step in and halt the construction by executive order, which is his prerogative as per the Supreme Court decision of 2000. Instead, he appointed, and the Supreme Court endorsed, an Oversight Group headed by V K Shunglu to investigate the status of rehabilitation in the valley. The Oversight Group, with the help of the National Sample Survey Organisation (NSSO) was to survey dam-affected villages in MP and report back to the court and prime minister. In the meantime, construction on the dam was allowed to continue.

Shunglu Committee

Hear No Evil, See No Evil

On July 11, two hours before the court would hear the matter, the Shunglu Committee finally released its report to the public along with the PM's announcement that he accepted its findings as "fairly accurate" and did not find any reason to stop dam construction. Among the committee's specific findings were that the state's estimation of PAFs was largely accurate; that out of 86 rehabilitation sites, 24 were poor and 25 average; that there were significant shortcomings in civic amenities and construction at these sites; and that there were 4,000 grievances pending by PAFs with the Grievance Redressal Authority (GRA). The committee asserted that cash compensation was an adequate substitute for land and that PAFs were accepting it voluntarily. Further, it recommended that clearance be given for construction up to the full dam height. Essentially, the Shunglu Committee's report, as will be shown, drastically underestimates the extent to which rehabilitation is incomplete, justifies illegal rehabilitation policies such as cash compensation, and refuses to acknowledge that even the shortcomings they did find in themselves make construction illegal. Before engaging the specifics of the report, however, it is important to put it into context.

The entire premise of the Oversight Group (or Shunglu Committee) was flawed from the beginning. The idea of allowing dam construction to continue while investigation of rehabilitation in the valley was underway, is entirely contrary to the law. In the face of overwhelming evidence that rehabilitation was incomplete, the Supreme Court and the PM were legally required to halt any further construction. Instead, they constituted the Shunglu Committee and gave it a mandate to report back to the PM on its findings by June 30. The PM then would have six days to consider the report and issue a recommendation to the Supreme Court, which would make a judgment by July 10. But, by this time, construction of the Sardar Sarovar up to 122 metres would be a *fait accompli*. The rains would have begun, and no matter what the committee found, up to 35,000 families would be facing the possibility of submergence in August-September, with or without rehabilitation. This, of course, is entirely illegal and a violation of the rights of the dam-affected families. And as past experience has shown, without the threat of halting dam construction, the respective governments have no incentive to actually provide rehabilitation to affected families. Any person can travel to the Narmada Valley and the various resettlement sites and see thousands of families affected at 110 metres and below, who have yet to receive land-based rehabilitation. But with no power to prevent submergence even if it did recognise this reality, the Shunglu Committee appeared to be merely a diversion from the start. One strongly suspects that its creation was simply a stalling tactic, allowing construction on the dam to proceed without prior rehabilitation.

Besides this fundamental flaw, the Shunglu Committee's methodology for conducting surveys in the Narmada Valley was itself highly problematic. This begins with the terms of reference given to it by the PM's office. First, the investigation only looked into R&R in Madhya Pradesh where 175 villages are affected at the 122 metre dam height, and not Maharashtra and Gujarat where 33 and 19 villages are affected respectively. While the governments of both states claim rehabilitation to be complete, this is far from the case with thousands of PAF sat 110m and below yet to receive cultivable, irrigable land and many still languishing without rehabilitation in the valley. But the Shunglu Committee did not even visit the villages in these states. Further, the committee only investigated those displaced at 110 and 122 metres, even though there are thousands affected beneath 110 metres in all three states who have yet to receive rehabilitation.

Further, the committee's investigation was hamstrung by its insistence on only using the Madhya Pradesh government's own data on PAFs, provided in its Action Taken Report (ATR). Thus, for the most part, the committee was only verifying whether the people on the government's existing PAF list were rehabilitated or not. While in some villages they took the names of those who claimed to be affected but were undeclared, people from multiple villages report that this was not done in their village. There was no comprehensive attempt to do complete village surveys and ascertain who in fact would be affected and was not on the PAF lists. Yet, one of the biggest problems with rehabilitation in the valley is that thousands of people have been excluded from the PAF rolls altogether. Many people are left off these lists because of faulty surveys, the failure of the government to recognise the traditional but untitled landholdings of adivasis, and the government's unwillingness to properly identify major sons of landholders (who are also entitled to land). In addition, over the past several years, the MP government has resorted to large-scale statistical manipulation in order to simply delete families from the PAF list whom it finds inconvenient to rehabilitate. An analysis of Narmada Control Authority (NCA) documents shows that GoMP has made terribly inconsistent estimates of PAFs, erasing thousands with the stroke of a pen from one statement to the next.

For example, at two meetings of the NCA's R&R Sub-Group (February and November 2002), GoMP testified that there were a total of 12,600 families from 135 villages in the state affected by the dam at 110m. This was the number it had consistently given at meetings over the course of several years. Nevertheless, in April 2003, it made the claim that now only 8,406 families were affected at 110m, thus reducing the number of PAFs it recognised by a third. The GoMP justified these reductions by making a novel distinction between "temporarily" and "permanently" affected people, arguing that only those affected permanently at the present height (meaning those families whose agricultural lands will be submerged for the entire year or whose house plot is affected) needed to be rehabilitated before a height increase. This is nonsensical since even if fields are submerged temporarily, the year's crops will be destroyed. Further, this distinction is contrary to the NWDTA and was found illegal by the Supreme Court in its 2005 decision. Nonetheless, these deleted PAFs are yet to be accounted for. So, by only verifying the government's existing ATR report, and not inquiring any further, the Shunglu Committee largely missed one of the biggest problems with rehabilitation in the Narmada Valley.

There are also problems with how the NSSO teams were actually asking the questions of those it chose to interview. The "surveys" (really just verifications of the ATR) were being conducted with standardised questions, demanding simple yes/no answers. With no open-ended questions, the teams could not delve into the complicated reality that people faced with respect to displacement and rehabilitation. This created numerous absurdities. For example, in one village an interviewee reported that he had not received compensation for his soon-to-be submerged house. Instead of noting this, the surveyor asked him if he had received a houseplot. The interviewee responded that he had on paper, but had not received compensation for his house and thus could not build a new one, and that moreover there were no civic amenities like electricity at the site. But the surveyor simply recorded that the person had received a houseplot! In another example, surveyors refused to listen to villagers who were trying to tell them that their land would become "tapu" (surrounded by water) because they did not have a column for that on their survey. Without open eyes and ears, how could the survey team understand the actual situation in the valley?

Further, reports from many villagers suggest inappropriate behaviour by government surveyors, which call into question their "independence" and impartiality. While surveyors initially stated that they would not be accompanied by government personnel, in some cases even NVDA officials were seen with the teams. This suggests improper influence by an agency that has a vested interest in continued dam construction and, as a part of the MP government, is itself a respondent in the Supreme Court case. Meanwhile, villagers not being directly surveyed and the people's representative organisation (the NBA) were not allowed to talk to surveyors. Even more disturbing, surveyors "propagandised" and even made speeches to villagers about the dam project. Some surveyors encouraged villagers to accept the meagre (and illegal) cash compensation that the government of Madhya Pradesh was offering instead of demanding the land they are legally entitled to. In another village, one of the surveyors (who happened to be from Gujarat) actually gave a speech to PAFs about how the dam was in the national interest and that they should make a sacrifice for it. Another told them that they should move to Gujarat. Thus, the NSSO teams did not at all appear to be conducting an impartial survey. Further, numerous reports of surveyors recording responses in pencil, and the refusal of the NSSO teams to distribute questionnaires to the people, have further undermined confidence in the survey.

Omissions and Inadequacies

Given these premises and methodology, the Shunglu Committee's report is suspected from the beginning. But, the report's specific findings and recommendations are also highly problematic, containing numerous underestimations and contradictions, and yet ironically still proving grave illegalities in rehabilitation. To begin, the report claims that the survey found "no deviations" between the GoMP's estimation of affected families and the survey data. However, it also reports that 6,485 families who are not on the PAF list claimed to be affected. But it then dismisses these claims on two grounds. First, it suggests ignoring 4,000 claims because they have cases pending before the GRA. This is nonsensical. These petitions to the GRA reflect the large-scale lack of rehabilitation in the valley; how can they be excluded without verification and, the conclusion reached that the government's PAF lists are correct? Second, the report claims that most of these petitions are bogus and made by people who have come to the villages and constructed houses for the purpose of receiving compensation, basing its argument on what it claims are dramatic changes in demographic census data. Yet this data shows clear absurdities and the allegations on which they are based, to be groundless. For example, the report's tables show the village of Picchodi and Anwli, among others, as having no population in 1991 and suddenly developing human habitation by 2001. Both are decades-old villages with approximately 800 and 600 families respectively, all PAFs. Meanwhile, the survey lists other villages such as Khujawa as having a population in 1991 and no population in 2001. Anyone can go to the village of Khujawa and see that there are approximately 125 houses with a population of roughly 900. Most glaringly, the Shunglu report itself shows that Khujawa villagers' houses were acquired in 2003. Whose houses were acquired if there was no population in the village? Thus, these few examples show the Shunglu Committee's "demographic changes," by which it is actually levying the charge of "cheating" on the people in the valley, are based on incorrect data and thus cannot be used to dismiss the over 6,000 people claiming to be affected.

Moreover, people in many villages such as Semalda of Dhar district, Kukra of Badwani district, and almost all of the 26 villages in Jhabua district claim that schedule II (the part of the survey that is supposed to enumerate people not on the official lists) was not even administered. This might be one of the reasons why the Shunglu report did not include a village-wise breakdown of the number of people claiming to be PAFs. In many of these villages, those present report that the surveyors spent minimal time in the villages, and in some cases didn't even get off their boats. In only a few villages, due to the presence of outsiders, did they conduct house-to-house surveys. This brings up a larger issue with the Shunglu report, which is how it is possible to do any kind of comprehensive survey of 177 villages in only 18 days. The joint task force conducted in Maharashtra, which surveyed only 33 villages affected by the dam, took a year and a half to complete. The Shunglu report itself admits that it was not possible to evaluate judiciously the petitions made by people claiming to be affected.

One of the most egregious aspects of the Shunglu report is that, in addition to simply verifying compliance with the law, it attempts to push its own interpretation of the law itself. Most problematically, it tries to advance the idea that cash compensation (known as the Special Rehabilitation Package (SRP)) is a legal substitute for land. The report states that, "SRP, in the opinion of GRA, is a legitimate substitute for providing land for land as stipulated in the NWDT Award." First of all, the GRA has no power to overrule the NWDTA and two Supreme Court decisions, which clearly mandate that rehabilitation must be land-based. The committee justifies its stance by saying that the acceptance of cash compensation was voluntary. But this is belied by its own finding that the land bank of Madhya Pradesh contains totally uncultivable and unirrigable land. Many of those who accepted cash compensation did so because the government refused to offer them good land, and they feared that they would get nothing else. Few have been able to purchase alternative land with the compensation they have received. The SRP is simply an illegal strategy for evading the responsibility to provide cultivable and irrigable land to every eligible family. Thus, the Shunglu Committee's advocacy and acceptance of cash compensation instead of land is unjustifiable and has no basis in law whatsoever.

Further, in its findings on the status of rehabilitation sites, the Shunglu report itself acknowledges their inadequacy and incompleteness. The report admits that 24 rehabilitation sites are "poor" and another 25 are "average." Leaving aside what these 25 "average" sites look like, this means that rehabilitation is not complete in at least 24 sites, which on average, are supposed to accommodate at least 200 families each. This means that even by the Shunglu Committee's generous count, approximately 5,000 families do not have a suitable rehabilitation site in which to reconstruct their lives and homes. In its description of these sites, the committee reports that, "Around 50 percent of the plots demarcated on government land appeared to be in an underdeveloped condition. These plots are demarcated either on the rocky areas or on slopes of hillocks and/or require substantial filling." Further, the report states that, "The water storage system at a majority of the sites was not satisfactory" and that "there is a need to ensure continuous supply of electricity." With regard to civic amenities, while some structures are built, "In almost all cases, these facilities are not functioning. These are just bare buildings without any provision of the staff and support structure for their functioning. Many of these buildings were constructed a few years ago and hence require repairs and proper maintenance. The absence of ponds and children parks is rather glaring." The report concludes that, "There are gaps between the R&R claims lodged by NVDA and those discovered." How then can the Shunglu

Committee recommend continuing construction when it itself finds such glaring inadequacies in the rehabilitation sites?

The Shunglu Committee report estimates that the deficiencies in the rehabilitation sites can be removed in the present financial year. But, the financial year has ended on March 31! So it is not even the three months that the committee suggests in other places (which is also illegal). March 31 is six months after submergence, not before as required by law. It is entirely illegal and unjustifiable for rehabilitation to be completed at such a late date. (And in fact, it is a conservative estimate: the GoM team reported that it would take at least a year, which is also an underestimation given the total absence of land-based rehabilitation in MP.) It is beyond human reasoning to understand how, given its own admission of such deficiencies and lagging rehabilitation, the Shunglu Committee can recommend that dam construction can continue.

Meanwhile, an alternative independent survey conducted by the *gramsabhas* of four villages with the help of outside experts, and a survey of eight villages done by an independent team of academics headed by Kamal Mitra Chenoy of JNU, both found large discrepancies between the government's rehabilitation claims and the ground reality. Both found many more people affected than reflected in the ATRs, highlighted numerous problems with the rehabilitation sites, and documented the failure of the government to offer alternative agricultural land to PAFs. The latter survey, coming on the heels of the NSSO surveys of the village, also uncovered numerous complaints by villagers about the inadequacy of the surveys. The report states, "In our sample villages and sites and in our interactions with hundreds of people in the Narmada Valley, we did not come across a single instance of people in the affected areas being asked to answer any rigorous sample survey/questionnaires designed specifically by NSSO. Instead PAPs were asked to respond to the ATR of their area that we reprepared by NVDA." Thus, the Shunglu report does not even appear to have attempted a serious, rigorous survey of villages in the Narmada Valley. Given its clear methodological shortcomings and contradictory findings, it is impossible to have faith in the Shunglu report. Further, given the fact that the report itself found substantial evidence of incomplete rehabilitation yet recommended continuing dam construction anyway, it is even more difficult to have faith in its commitment to verify and ensure the fair and legal rehabilitation of affected families.

For the people of the Narmada Valley, the Shunglu Committee and the inaction of the Supreme Court and the PM are just the latest in a long series of shams and betrayals. Countless committees and investigations have come and gone. The independent investigation of the World Bank found the project to be fatally flawed, and led to the World Bank pulling out of the project. Yet, the Indian government, prompted along by Gujarat, has persistently pushed the project forward, regardless of the law or its human consequences. In the face of verifiable facts to the contrary, the governments have perpetually claimed rehabilitation to be complete. When one is in an actual village in the Narmada Valley seeing first hand evidence that disproves such claims, one can only be baffled that such lies can be accepted at the highest levels of government. The French philosopher Michel Foucault famously posited the inextricable relationship between truth and power. The Narmada saga is a perfect illustration of how the powerful can manufacture demonstrably false "truths" to obscure their real actions and further their own interests.

Conclusion: On the Impossibility of Rehabilitation

It is also important, however, to step back from the legal intricacies and details of the SSP and recognise its larger lesson. Large dams are emblematic of a development paradigm characterised by giganticism, the fetishisation of technology, and an extreme faith in the capacity of technical knowledge to transform the world. But, as the SSP shows, the social and environmental consequences of meddling on such a large scale are exceedingly complex, and very often surpass the ability of governments to cope with. When a state tries to uproot existing communities and remake them elsewhere, it fails because the social, economic, and ecological relationships that characterise any village, much less hundreds of villages, cannot be adequately understood (and therefore replicated) from above. In a situation characterised by resource scarcity (there is little excess land in India), and a corrupt bureaucracy, the task of large-scale rehabilitation becomes even more difficult. When a development model that requires large-scale displacement is coupled with a government that fails to respect the legal and human rights of those being sacrificed for "the national good," the situation becomes downright dangerous.

If anything, the failure of the Indian government to adequately rehabilitate the displaced people of the Narmada Valley shows the failure of human institutions to operate on this scale. While it must be emphasised that the government is not doing a fraction of what it could to rehabilitate PAFs, it is also clear, that full rehabilitation is impossible. To even provide the legally required rehabilitation of two hectares of cultivable, irrigable land, and community resettlement sites with civic amenities, is an impossible undertaking for half a million people. There is simply no vacant land in India to do this. To actually restore people to the standard of living they enjoyed before displacement is even more impossible. Adivasis of the Narmada Valley not only possess fertile cultivable land, but also have access to common grazing lands, forest, and river, which provide a crucial part of their livelihoods. Even in the few places where PAFs have received some land, these other subsistence resources have been absent. As a result, people must buy what they used to get for free in their villages. Their cash requirement thus increases, making subsistence farming unsustainable. Cash farming with such small holdings is not remunerative enough in itself, especially on bad land. As a result, most displaced people have become proletarianised and thrown into the precarious existence of day labourers. This is an example of the complex chain of consequences that accompanies large-scale attempts to displace and resettle people. The Indian government cannot hope to manage the consequences of doing this to hundreds of thousands of people. It is this impossibility that led the NBA to outrightly oppose dam construction in the first place.

But what the NBA perceived from the outset has failed to deter the Indian government. As a result, up to 35,000 families will soon be displaced from their homes, communities, and resource base and shoved into make-shift rehabilitation sites with no adequate means of livelihood. This entirely illegal and inhuman situation is the product of a flawed development paradigm coupled with a government that at its highest levels refuses to respect the law and human rights. What is at stake in the Narmada Valley is not just the lives of five lakh people, but the very nature of democracy and the meaning of development. One suspects that there might just be some truth to the NBA slogan, 'Narmada Bachao, Manav Bachao!'.

(Mike Levien, is a Graduate Student in Sociology at the University of California- Berkeley. He spent two years working with grassroots community development initiatives in the US and Central America. He then came to India and is working with the Narmada Bachao Andolan. He can be reached at mlevien@berkeley.edu).

8

Spotlight on Indus River Diplomacy: India, Pakistan and Baglihar Dam Dispute

Robert G Wirsing and Christopher Jasparro

According to the Indus Waters Treaty (IWT), India and Pakistan are convinced to the permanent partitioning of the Indus river system. India has free rights for the waters of the three eastern rivers (Ravi, Beas, Sutlej), and Pakistan has ownership of the waters of three western rivers (Chenab, Jhelum, Indus). This article talks about the issues raised by Pakistanis on the design of Baglihar dam on Chenab River in Jammu and Kashmir. The author says that this rivalry over river water may increase the tension between India and Pakistan.

Overview

- India is moving steadily closer to a danger zone in terms of water supply. In the last 50 years per capita availability of water in India has declined by roughly 60 percent with an equally precipitous drop possible in the next 50 years. Meanwhile, Pakistan may be nearing "water stress" limit of 1000 cubic meters per person per year, below which serious economic and social consequences are likely.

- Rivalry over river resources has been a chronic source of severe interstate tension between India and Pakistan. With river resource issues intensifying, the possibility for violent interstate conflict will likely increase.
- Even if direct violence is avoided, inability to resolve river resource issues will undoubtedly limit the ability of both countries to manage and utilize water resources in the most efficient manner. Inadequate management of water resources will exacerbate domestic problems in these demographically explosive societies, which could lead to a variety of unwanted conditions such as increasingly fertile grounds for political extremism and terrorism.
- The outcome of the important on-going dispute over the Baglihar dam has broader implications not only for future management of increasingly important interstate river issues between India and Pakistan and in the entire region of South Asia, but also for the overall character of future India-Pakistan relations. Given the current differences between the parties, the prognosis is not encouraging.

When speaking of the river resource issue in India-Pakistan relations, one is strongly tempted to caption it cleverly with the title of Norman Maclean's poignant novella, *A River Runs Through It.*[1] For the Indus river runs through the history of India-Pakistan relations every bit as consequentially as the Big Blackfoot River ran through the lives of Maclean's fictional Montana family. By the same token, Maclean's solemn avowal at the end of the piece—"I am haunted by waters"—might well have occurred to more than a few Indian and Pakistani diplomats over the years as they contemplated the river-ensnarled diplomatic agenda between the two co-riparian neighbors. The national destinies of India and Pakistan are inextricably joined by the Indus river whose waters they share; and no aspect of their diplomatic relationship bears more heavily on those destinies than that pertaining to the Indus river.

For over a half century, bitter rivalry over river resources has been a chronic source of severe interstate tension between India and Pakistan. It has arguably been one of the leading causes of full-scale warfare between them. Today India and Pakistan are both faced with rapidly escalating problems of acute river resource scarcity. Intensified rivalries over river resources could precipitate violent interstate conflict between them in the future. Even if direct violence is avoided, the inability to resolve river resource issues between them will undoubtedly limit the ability of both countries to manage and utilize water resources in the most efficient manner. That, in turn, can be counted on to exacerbate intrastate resource conflicts as well as related domestic disaffection—developments that appear certain to reduce even further the Indian and Pakistani states' capacities to manage peacefully the interstate resource disputes between them. Increased tensions over water, in other words, help exacerbate or intensify overall tensions, thus at worst creating a more favorable environment for interstate conflict or at least making resolution of interstate security issues between the countries even more difficult.

The outcome of the important on-going dispute over the Baglihar dam, discussed in some detail below, can serve as a bellwether both as to whether India and Pakistan will be able to manage increasingly important interstate river issues (and hence insure internal stability and development) and also as to what direction future overall relations between these countries is likely to take.

A distinguishing characteristic of the Indo-Pakistan river resource relationship is that, in sharp contrast with India's river relationships with its other major co-riparians in the region, Bangladesh and Nepal, there is a comprehensive treaty—the 1960 Indus Waters Treaty (IWT)—that was deliberately designed to settle with one stroke, and permanently, the matter of water sharing. The IWT accomplished this by getting India and Pakistan to consent to the permanent partitioning of the Indus River system—India winning unfettered ownership of the waters of the three eastern rivers (Ravi, Beas, Sutlej), and Pakistan acquiring nearly unfettered ownership of the waters of the three western rivers (Chenab, Jhelum, Indus). Often cited as the only major bilateral agreement between India and Pakistan to have stood the test of time, the IWT is today coming under extraordinarily close, in some cases highly critical, scrutiny. There are observers on both sides of the border, and representing opposites points on the political compass, who complain that the treaty is out of date, that it obstructs rational exploitation of the Indus river's resources, and that it ought at least to be amended, if not entirely scrapped.

One reason for dissatisfaction with the IWT is that, as presently constructed, it offers very thin support to the integrated or joint development of the Indus river basin. After all, the treaty's success, in the face of huge distrust and animosity between the two signatories, had largely to do with its abandonment of customary international norms governing internationally shared rivers. In particular, it discarded the norms protecting the downstream country's traditional uses of the river waters, in place of which it offered geo-physical partition of the river system itself. This formula was conceivable only in the unique geographic and political circumstances of the Indus basin. The *division of the waters,* in its own way, represented the "unfinished business" of the subcontinent's 1947 *territorial* division. In the judgment of B G Verghese, one of India's most frequent commentators on river resource issues and an advocate of "joint investment, construction, management and control" of the three western rivers "allocated to Pakistan but . . . under Indian control," Article VII (on Future Cooperation) and Article XII (a provision allowing for agreed modification of the treaty) provide ample license for constructing what he calls an Indus-II "on the foundations of Indus-I." Indus-II, he says, should "be fed into the current peace process as a means both of defusing current political strains over Indus-I and insuring against climate change. It could reinforce the basis for a lasting solution to the J&K (Jammu & Kashmir) question by helping transform relationships across the LoC (Line of Control) and reinventing it as a bridge rather than merely as a boundary-in-the-making."[2] Verghese's colleague and longtime collaborator at New Delhi's Centre for Policy Research, Ramaswamy Iyer, disagrees. According to him, the existing IWT is poorly designed for the kind of Indus II that Verghese proposes. The IWT, he argues, "was a negative, *partitioning* treaty, a coda to the partitioning of the land. How can we build cooperation on that basis?" If a new relationship between the two countries on the Indus is desired, and Iyer indicates he is in complete agreement with Verghese on that, then "a totally new treaty will have to be negotiated; it cannot grow out of the existing treaty. . . ." That undertaking, he suggests, would face enormous complications. "Perhaps," he concludes, "it would be better to leave things as they are, and hope that with improving political relations a more reasonable and constructive spirit will prevail in the future than in the past."[3]

A second reason for dissatisfaction with the IWT is that, in practice, the treaty favors one side over the other. Pakistanis hold that they gave up more water than they gained, that the diversion of Indus river waters required to compensate for the loss to India of the three eastern rivers that has inflicted heavy ecological penalties upon Pakistan, and that—worst of all—India's retention of the right to "non-consumptive" uses of the three western rivers presents Pakistan with the endlessly frustrating and ultimately futile task of guarding its water resources against Indian poaching. Indians, in turn, hold that it is *their* side that gave up too much water in the 1960 treaty, and, moreover, that Pakistan has made it virtually impossible for India to exploit effectively the non-consumptive uses, the production of hydropower in particular, allowed them on the western rivers.[4] Nothing better illustrates this dimension of the debate over the IWT than the current diplomatic wrangling between the two countries over the Baglihar dam.

Diplomacy of the Baglihar Dam[5]

The Baglihar hydropower dam is located on the Chenab River in Doda district about 110 kilometers eastward of the Pakistan border in the Jammu division of the Indian state of Jammu and Kashmir. Currently, construction is said to be somewhere between one-third and one-half complete. The dam, when finished, will rise to 144.5 meters and have an installed capacity of 450 MW (900 MW, when a second phase power station is built). The Baglihar is one of that eleven reported major hydroelectric projects that India has identified in Jammu and Kashmir, nine of them on the Chenab.[6] Along with two others, the Wullar dam (officially labeled by Indians the Tulbul navigation project) and Kishenganga hydropower project, the Baglihar dam project is presently the focus of intense diplomacy between India and Pakistan. Two rounds of formal bilateral talks on the Baglihar, first in June 2004 and then in early January 2005, failed to reconcile their positions on the question of the dam's conformity to the restrictions set forth in the treaty's extraordinarily detailed annexes and appendices. In mid-January 2005, Islamabad invoked the arbitration provisions of the IWT, the first time this had happened in the treaty's history, and requested the World Bank, the formal "guarantor" of the treaty, to appoint a neutral expert.[7] A Swiss civil engineer, Raymond Lafitte, was appointed in May 2005, visited the site of the dam in early October 2005, and was expected to submit his report—the findings of which, in accord with the treaty, would be final and binding on both sides—sometime in the summer of 2006.[8]

The dispute over the Baglihar is technically complex. To simplify a bit, the Pakistanis have raised three key sets of technical objections to the design of the Baglihar dam. One set of objections relates to the dam's *storage* capacity, a second to the *power intake tunnels*, and a third to the *spillways.* As for the dam's storage capacity, Pakistani officials call attention to the treaty's allowance of only "run of the river" dams. Such dams are by definition *non-storage* dams—in other words, power is generated from normal river flow, the tapping of running not dammed water. In practice, Pakistanis concede, some storage is essential (and is explicitly authorized by the treaty): there is, after all, considerable (especially seasonal) variation in the flow of rivers, a fact that necessitates installation of sufficient storage to enable stable, efficient operation of the hydroelectric plant on a regular, year-round basis. What Pakistanis object to, on the one hand, is the 144.5 meter height of the Baglihar dam, which they say exceeds by nearly 100 meters what they are prepared to accept as a run of the river project and, on the other hand, the size of the live storage, or pondage, that a dam of this height allows. According to the Pakistanis, the pondage area at Baglihar is not consistent with the treaty, which was concerned above all, as they see it, with preventing India from exerting control over the western rivers' flow.

As for the power intake tunnels, the Pakistanis object to the fact that there are two of them, which they say is not permitted by the treaty, and to their position—not high enough according to the Pakistanis, who once again want to minimize Indian discretion when it comes to the discharge of waters. The higher the power intake tunnels, the lesser the opportunity for them to be used to release large quantities of stored water.

Much the same reasoning accounts for Pakistani objections to the design of the spillways. Baglihar's spillways are gated, which Pakistanis have argued was unnecessary, and the gates, say the Pakistanis, reach lower (32 meters below the effective top of the dam) than they should. Once again, the issue is one of Indian control of the stored waters: ungated spillways or shorter spillways, as Pakistanis see it, are more in conformity with the treaty.

Pakistani officials maintain that the Baglihar dam's design supplies India with the means, on the one hand, to economically squeeze, starve or strangulate Pakistan, or, on the other hand, to flood Pakistan, conceivably for military purposes. They argue, moreover, that the Baglihar dam has huge precedent-setting importance: for Pakistan to compromise on Baglihar, they say, would set a precedent that India could invoke whenever it liked elsewhere on the Chenab or Jhelum rivers. A trickle of Pakistani deviations from the treaty today, said one senior Foreign Ministry official, could become a flood of them tomorrow.

Pakistani officials also cite the Baglihar's political importance. It is, they concede, an extremely sensitive domestic political issue. The government's political foes demand to know why it took so long to refer the matter to the World Bank. "Baglihar is a politically painful matter for Islamabad," admitted one Pakistani official. Baglihar and other Indian hydroelectric projects, say the Pakistanis, are also extremely useful tools which New Delhi uses to win the political support of energy-deficient Kashmiris—and to drive a wedge between Kashmiris and Pakistanis. Additionally, Pakistanis point out, the Baglihar case effectively tests the thus-far-untested arbitration mechanism in the treaty's Article IX. The treaty is Pakistan's lifeline to the waters of the western rivers. The likelihood exists that its arbitration provisions will be invoked much more often in the future.

Asked to characterize Indian negotiating strategy, Pakistani officials asserted that (i) it was one of delay, of foot-dragging, of "tiring you out"; (ii) of "creating facts"—proceeding with construction plans, even when aware that the plans might well violate the treaty, so that Pakistan, confronted eventually with *fait accompli*, would have no choice but to cut its losses and accept an unfavorable compromise settlement; and (iii) insisting on a bilateral framework of talks, without intending ever to settle on any but India's terms.

Indian officials naturally have a rather different "take" on Baglihar. Its design, they contend, is fully in compliance with the treaty. Notwithstanding Pakistani objections, Baglihar, according to them, is a run of the river dam. India has built nearly 20 such dams, they point out, and neither the Baglihar's height nor storage capacity disqualify it for designation in this category. The Pakistanis, say Indian officials, deliberately obstruct and willfully interpret the treaty in an excessively restrictive manner. Their raising of objections to Indian projects, they say, is compulsive and ritualistic, and not based on impartial assessment of the facts. In truth, they say, the treaty's language is quite flexible, allowing adjustments that take advantage of modern dam engineering technologies. Pakistani objection to the positioning of the power intake tunnels, for instance, ignores the treaty provision specifying that they should be constructed at the highest level *consistent with sound engineering*. Sound engineering, they say, requires construction of intake tunnels to maximize the "water seal"—the elimination of air from the tunnel, which was an important element that went into the design decision. Likewise, Pakistani objection to the gated spillway was "absurd". Himalayan rivers, a senior Indian official pointed out, carry enormous quantities of silt, far more than one generally finds in rivers in the West. Gated spillways, lower positioned gated spillways in particular, are essential to flush the silt-laden waters through the dam. Otherwise, the silt bombards the wall of the dam, falls to the bottom and swiftly builds up sediment on the river floor—a development that modern dam builders seek to thwart in order to prolong the useful life of the dam. In the view of Indian officials, the treaty authors could not possibly have intended that hydropower projects built in 2005 should be designed to conform to technologies in use in the 1950s.

Indians also argue that Pakistani anxieties about India's acquiring the ability to shut off the flow of water downstream, posing a threat to the economically vital farmlands of Punjab, have no basis in reality. Pakistanis claim that once its live storage had been drained off, refilling the Baglihar in the dry season would take a full 26 days, giving Indians ample time to disable the Punjab's largely river-fed irrigation system. The Indians, however, contend that the process of refilling would take no more than 19 days—not enough time, as they see it, to throttle their neighbor's agricultural economy!

Two reasons for dissatisfaction with the IWT have been considered here: first, that as a postscript to the region's territorial partition, it offers very thin support to the integrated or joint development of the Indus river basin; and second, that the treaty, in practice, favors either one side or the other. In the case of the Baglihar dam, for instance, Indians have ineluctably been led to view the treaty mainly as an impediment to be artfully bypassed in the drive for increased hydroelectric power.

These two reasons are brought together with a third, still more disturbing, reason in an unusually provocative book, *The Final Settlement: Restructuring India-Pakistan Relations*, brought out by the Mumbai-based Strategic Foresight Group in 2005. This third reason for dissatisfaction is that the treaty, though highly unlikely to be abrogated by India, offers only a very frail defense against heightened conflict over river resources between India and Pakistan, and that it is only a matter of time before water war becomes a virtually unavoidable feature of the region's political environment. In a chapter entitled "Water" and with the subtitle "The Secret," *The Final Settlement* holds that water has been central to the Kashmir dispute from the beginning, that the public debate over Kashmir—focused on lofty goals of self-determination and human rights (and not on Islamabad's self-interest in water security)—has always been discreetly steered away from this fundamental fact,[9] and that Pakistan's mounting water insecurity virtually ensures a still deeper and volatile nexus between water and Kashmir in coming years. The book cites as evidence, frequent unofficial Pakistani expressions of interest in recent years in a so-called Chenab formula of conflict resolution, according to which Jammu and Kashmir would be further partitioned, with Pakistan being granted the Kashmir Valley *and* a substantial (and Muslim majority) portion of Jammu, enough to give it command of the Chenab River. The Chenab, in *The Final Settlement*'s view, is the ultimate prize, possession of which by Pakistan would virtually end its water woes: with the 1960 treaty effectively terminated, Pakistan would be able to develop the Chenab's potential to the maximum, not only in terms of storage dams for irrigation but also for hydroelectric power and flood control. This, according to the book, has in recent years been the latent objective of Pakistani diplomatic and political activity relating to Kashmir.[10]

Most disturbing, from *The Final Settlement*'s perspective, is that what Pakistanis feel they must have, Indians will never give up. The Chenab river is clearly not for sale. This could have dire consequences. "The treaty," according to *The Final Settlement*, has engendered a vicious cycle. Lack of trust between India and Pakistan forced the bifurcation of the Indus River Basin. As the gap between water availability and requirements widens in Pakistan, its desire to intensify jihadi operations will grow. Agricultural development will be affected, which, in turn, will produce a stratum of unemployed youth willing to service terrorist groups. This, in turn, would aggravate the mistrust and hostility between the two countries. This vicious cycle of depleting resources spawning unemployment and fueling terrorism is feared to intensify in the near future.[11]

The oversimplified correspondence between water availability and terrorism in *The Final Settlement*'s analysis should trigger some skepticism about all its arguments. It is not alone, however, in calling attention to the potentially severe security implications of the region's water resource rivalry. Echoing some of the gloom implicit in the above quotation, a senior Pakistani diplomat told one of the authors of this article, "Water has become *the* core issue between India and Pakistan. . . . (As a result,) India-Pakistan relations will retain (in the future) the same level of tension (as they now have)."[12]

Using words reminiscent of the appeals by Verghese and others for greater India-Pakistan collaboration in the development of the Indus basin's water resources, *The Final Settlement* does finally end on a positive note. "An alternative approach to the Indus treaty issue," it says,

> *could be an integrated development plan for the conservation of the Indus Basin. The plan, to be jointly developed by India and Pakistan, would involve a creative solution to the political dimension of the conflict in Jammu & Kashmir. It is imperative for both India and Pakistan to envisage comprehensive development and planning in the (Indus) River Basin. A holistic approach to water resources—recognizing the interaction and economic linkages between water, land, the users, the environment and infrastructure—is necessary to evade the impending water crisis in the subcontinent. . . . The integrated development approach is Utopian. It is only possible with a paradigm shift in mindset and complete end to hostilities, both physical and psychological.*[13]

To say the least, the recommended "integrated development approach" offers very meager encouragement. One must wonder if it is, in fact, a viable alternative. Should it not prove viable, it likely bodes ill for the future of India-Pakistan relations. An integrated or basin approach to water resource management may well turn out in the future, in the face of mounting water scarcity in the region, to be not only an attractive but *essential* approach to water resource management—in the absence of which domestic problems in these demographically explosive societies could well themselves become unmanageable, an outcome that could lead to a variety of unwanted conditions, increasingly fertile grounds for political extremism and terrorism among them. On the other hand, should the bilateral relationship progress to the point where an integrated development approach becomes possible, then perhaps vicious cycles of deepening internal dismay and escalating interstate tension can be halted or even reversed.

(Robert G Wirsing is Professor at Asia-Pacific Center for Security Studies (APCSS) in Honolulu, Hawaii. His area of expertise is South Asia and Identity Politics. Earlier he was Professor of International Studies at University of Carolina.

Christopher Jasparro is a researcher at the APCSS. He is a political and cultural geographer with additional training and professional experience in transportation systems management).

Endnotes

1 Norman Maclean, *A River Runs Through It and Other Stories* (Chicago: University of Chicago Press, 1976).

2 B G Verghese, "It's Time for Indus-II," *The Tribune* (Chandigarh), 25 May 2005.

3 R R Iyer, "Indus Treaty: A Different View," *Economic and Political Weekly*, vol.11, no. 29 (16 July 2005), p.3144.

4 According to Verghese, on virtually every one of the 27 occasions since signing of the IWT when India has passed information to Pakistan, in accord with treaty provisions, on planned withdrawals or construction on the western rivers, Pakistan has raised objections. In his view, "the objective has been political and the motivation to delay if not deny progress that primarily benefits J&K" B G Verghese, "Fuss Over Indus-I: India's Rights are Set Out in the Treaty," *The Tribune* (Chandigarh), 25 May 2005.

5 This section relies in part on interviews with Indian and Pakistani government officials, including key participants in the Baglihar talks, conducted by one of this article's authors in Islamabad and New Delhi in April and September 2005. The interviews were generally conducted on the basis of non-attribution.

6 S Waslekar, *The Final Settlement: Restructuring India-Pakistan Relations* (Mumbai: Strategic Foresight Group, 2005), p.58.

7 The IWT provides for multiple levels of conflict resolution. At the start-up of any project it plans to construct on the western rivers, India is required to provide Pakistan with advance notice, including detailed plans and design. Pakistan has the right to raise questions about any aspect of the project. The "question" may be settled at either the level of the Indus River Commission, a treaty-authorized body consisting of two commissioners, one appointed by

Commission, a treaty-authorized body consisting of two commissioners, one appointed by each side, or at a higher-level inter-governmental meeting. Failing agreement at that level, a "difference" is said to exist (the situation that now obtains in regard to the Baglihar), a condition warranting the World Bank's appointment of a neutral expert. The neutral expert's task is strictly to determine whether or not the project design conforms to the treaty provisions. The next level, at which a "dispute" is acknowledged to exist, would require appointment by the World Bank of a Court of Arbitration. As the treaty's guarantor, the World Bank's role is that of go-between; it does not have any enforcement powers.

8 It was recently reported that the Swiss expert, Raymond Lafitte, had called Pakistani and Indian representatives for a final hearing on the Baglihar issue in the last week of May 2006. His final report was expected to be released soon thereafter. Ahmad Fraz Khan, "Baglihar Dam Final Hearing Next Month," *Dawn* online edition, 14 April 2006.

9 "To the outside world," the book observes, "it is projected that Pakistan is supporting a struggle for self-determination for the people of Kashmir. Within the closed-door precincts of General Head Quarters in Rawalpindi, Kashmir has a different meaning." Waslekar, *The Final Settlement*, p.59.

10 Waslekar, *The Final Settlement*, pp. 47-53, 73-78.

11 Waslekar, *The Final Settlement*, p. 68.

12 Interview, Islamabad, 7 April 2005. Identity withheld on request.

13 Waslekar, *The Final Settlement*, p.79.

Baglihar Dam Verdict

On 12th February 2007 Swiss Neutral Expert Raymond Lafitte, appointed by the World Bank, gave the final decision on 16-year-old dispute between India and Pakistan over the 450 MW Baglihar dam proposed on the Chenab River in Jammu & Kashmir. He delivered this report in presence of diplomats from both India and Pakistan in Berne, Switzerland. It took around two years to come to a decision which is a win-win situation for both the countries. The World Bank has framed it in such a way that both countries don't get any reason to object.

Pakistan had objected on three issues – water storage capacity, placement of spillway gates and the water intake – in the design of the Baglihar dam. It said that these design elements are against the Indus Water Treaty (IWT) signed in 1960. It feared that the extra height of the dam might allow India to store more water. In 2005 it requested the World Bank to intervene in this issue.

The judgment asks India to reduce the freeboard (height of the dam) by 1.5 meters. According to India this reduction does not affect the electricity production capacity of 450 MW. It also permits India to build run-of-the-river hydropower projects on the western rivers.

Lafitte approved India's calculation of flow of 16,500 cubic metres of water per second (cumec) into the dam as against 14,900 cumec figured by Pakistan considering the climate change. He totally rejected Pakistan's calculation of 6.22 million cubic metres (MCM) of pondage (amount of water held in the dam) and arrived at 32.56 MCM nearer to India's calculation of 37.72 MCM.

Lafitte also favored the construction of gated spillways (openings built at the bottom of dam discharge the sediment) after a detailed study of 13000.dams across the world and looking to the conditions at the dam site. He rejected Pakistan's objection saying that today's advanced technology, which was not available in 1960, can be used to manage heavy sedimentation that reduce life of plants.

The third issue was the water intake by turbines. According to treaty, the turbines must be located at the highest level. With the increase in height of turbines, the water quantity which can be released decreases. That is why Pakistan wanted 7m increase in the turbine positions but Lefitte allowed only 2m increment.

These changes slightly increase the project's costs but the judgment will facilitate India to complete the project by the end of December 2007. G Parthasarathy, former High Commissioner to Pakistan, said that with this verdict Kashmiri nationalist politicians will understand that India wants their development, but the Pakistanis believe in delaying the development process.

Complied by Shashikala Choudhary, Research Associate, Icfai Business Schnool Research Centre, Ahmedabad.

9

Muted Elation

T S Subramanian

On February 5, 2007, the Cauvery Water Disputes Tribunal declared its final award on the sharing of river water by Tamil Nadu, Karnataka, Kerala and Puducherry. This verdict has probably brought this issue to rest. For more than 40 years, there has been an unstable situation in the delta in Tamil Nadu regarding river water sharing. The author says there was a sense of elation and justice in Tamil Nadu that it is going to receive 192 tmc ft of Cauvery water from Karnataka at Biligundlu area.

There is cautious optimism in Tamil Nadu over the final award despite conflicting views being aired by political parties.

It was Friday, February 2, and there was a quiet sense of optimism in the Cauvery Technical Cell office in an old bungalow at Egmoré in Chennai. As if to confirm the optimism, two men were painting a bust of the moustachioed Colonel WM Ellis in front of the office. One of them carefully painted the words on the pedestal: "Col. WM Ellis, CIE, Royal Engineer, Architect of Mettur dam, River Cauvery..." The painting of the bust, after long neglect, was not without significance. On February 5, the Cauvery Water Disputes Tribunal was to

Source: Frontline, Volume 24, Issue 3, Febrauary 10-23, 2007.

pronounce its final award on the sharing of the river water by Tamil Nadu, Karnataka, Kerala and Puducherry. If the Krishnarajasagar dam symbolises the aspirations of farmers in Karnataka, its counterpart in Tamil Nadu is the Mettur dam, serving as the lifeline for thousands of farmers in the Cauvery delta districts in the State.

When the final award was announced and it became known that Tamil Nadu would receive 192 tmc ft of Cauvery water from Karnataka at Biligundlu, there was a sense of elation, a feeling that the State had received justice. But there were no celebrations.

The reaction of Chief Minister M Karunanidhi, who has handled the sensitive issue for the past 38 years in different capacities, first as Public Works Minister and then during five terms as Chief Minister, was circumspect. "The justice we have got after a long time gives us consolation," he said. When asked whether the final award gave him "full satisfaction", the 82-year-old Karunanidhi's reply was: "I will be fully happy only when the people of Tamil Nadu and Karnataka unitedly accept this award."

On the final award, he said: "It comes as a consolation because we have got justice over a demand we have been making for a long time." On the proposed "bandh" in Karnataka, he had a ready pun: "I want the Tamilians in Tamil Nadu and the Kannadigas in Karnataka to be "bandhus" [kin]. Hence I appeal to them not to organise the bandh," he said. He recalled that he had held negotiations with 11 Karnataka Chief Ministers since 1968 to break the deadlock, and paid a tribute to former Prime Minister VP Singh who set up the Tribunal in 1990.

Political parties and farmers' organisations said the absence of a "distress-sharing formula" in the award was "a serious flaw", though the final award did say that in years of lower rainfall the four riparian States would "proportionately reduce" their shares of the waters. There were important reasons for the disappointment at the absence of a more specific "distress-sharing formula". It was the absence of such a formula that allowed Karnataka not to implement the interim award in the years of "distress" (when rainfall was low) between 1991 and 2006. This led to Tamil Nadu farmers losing either the "kuruvai" or the "samba" crops, or both, in the years when both the southwest monsoon and the northeast monsoon failed.

K Balakrishnan, general secretary of the Tamil Nadu Kisan Sabha, affiliated to the Communist Party of India (Marxist), said that because the interim award did not specify a distress-sharing formula, "there was no way out" of the tangled skein that the issue had become. He insisted that the proposed Cauvery Management Board should be given the power to implement any distress-sharing formula that might be devised later. "The Tamil Nadu government and political parties in the State should constantly pressure the Centre to force Karnataka to implement the award and a distress-sharing formula that may be arrived at in future," said Balakrishnan. Inter-State water disputes became too complicated if they were allowed to persist, he said, adding that the Centre should make Karnataka obey the final order. He said that 192 tmc ft was "an acceptable share" for Tamil Nadu. The Tribunal had arrived at the final award only after studying extensively the documents supplied by the four States and listening to the arguments of their counsel. There was no other forum that could render better justice.

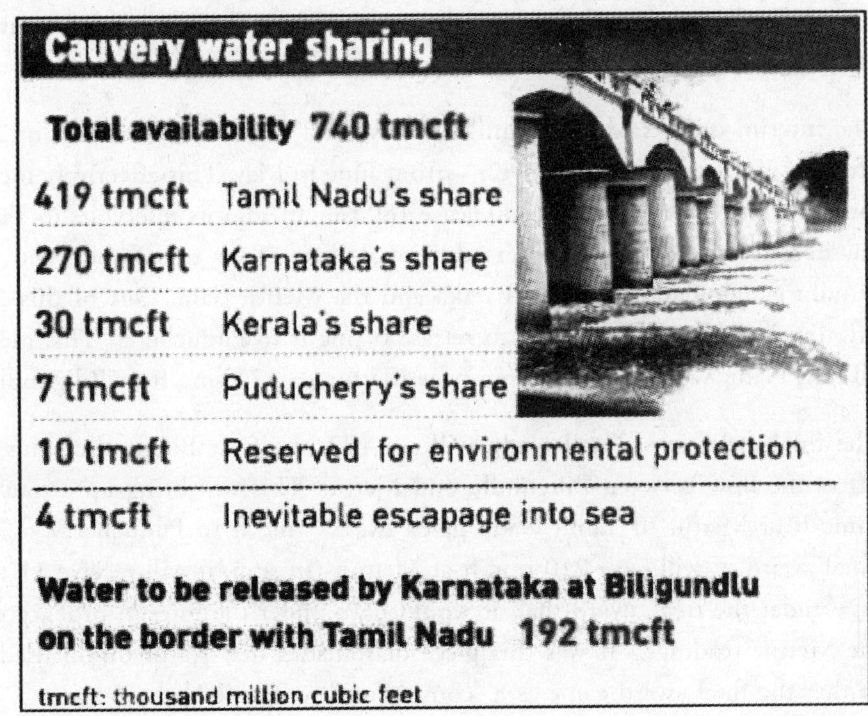

The final award has generated an expectation among the farming community in the State that it would set at rest a dispute that has defied a solution for so long. Balakrishnan said, "For more than 40 years, there has been an unstable situation in the delta in Tamil Nadu. Farmers could not plan their operations with any guarantee. Now that the final award has been pronounced, we hope farmers will have an assured supply of water."

S Ranganathan, general secretary of the Cauvery Delta Farmers' Welfare Association, the first organisation to file a case in the Supreme Court in 1983 for the constitution of a Tribunal, said the final award should have been "more specific" about what would be done in years of deficit rainfall.

The final award, coming after a long-drawn, bitter dispute, has renewed the demand for nationalisation of inter-State rivers, interlinking of rivers in the southern peninsula, and the 'lining' of canals. Financial institutions were unwilling to provide loans to the Tamil Nadu government for the lining of canals in the delta because of the uncertain situation that resulted from the non-implementation of the interim order by Karnataka.

The interim order said that Tamil Nadu should get at the Mettur dam 205 tmc ft from Karnataka in a water year – from June to May. This effectively meant that it was enough for Karnataka to release 180 tmc ft from its reservoirs to Tamil Nadu, because there would be an additional flow of 25 tmc ft between the Biligundlu gauging station in Karnataka and the Mettur dam. Out of this 205 tmc ft, Tamil Nadu was supposed to release 6 tmc ft to Puducherry. This meant that Tamil Nadu would get 199 tmc ft at Mettur or 174 tmc ft at Biligundlu.

The final order says Tamil Nadu will get 192 tmc ft at Biligundlu, plus 25 tmc ft as the flow between Biligundlu and Mettur. This will bring up a total of 217 tmc ft at Mettur. If Tamil Nadu gives away 7 tmc ft to Puducherry, as per the final award, it will have 210 tmc ft at Mettur. (In sum, it will receive 11 tmc ft extra under the final award than it would have under the interim order, going by the Mettur readings.) It was this piece of statistics that Karunanidhi used to argue that the final award came as a "consolation" to Tamil Nadu.

By and large, political parties reacted on predictable lines, depending on whether they were allies of the ruling Dravida Munnetra Kazhagam (DMK) or its rival, using the figures in the final award either to argue that the State had received justice or to establish that it had been betrayed.

While welcoming the final award, the CPI(M) and the Communist Party of India (CPI) underscored the need to ensure that Karnataka implemented it. N Varadarajan, CPI(M) State secretary, was confident that the final award would provide a permanent solution to the dispute. The Centre and the State governments should set up the necessary administrative mechanisms to ensure not only that Tamil Nadu got 192 tmc ft at Biligundlu in a normal year, but also that it got its due share in a year of deficit, Varadarajan said. D Pandian, CPI State secretary, said that the Centre should immediately set up a Cauvery Management Board to implement the final award.

The Pattali Makkal Katchi (PMK), though an ally of the ruling DMK, opposed the award. PMK founder S Ramadoss said the award disappointed the people of Tamil Nadu: he wanted the DMK to convene an all-party meeting to decide what should be done about it. Thol. Thirumavalavan, leader of the Dalit Panthers of India, another ally of the DMK, hedged his bets. He said the final award did not quite "do justice to Tamilians" but was a "consolation" to them. L Ganesan of the Bharatiya Janata Party (BJP) welcomed the award.

The All India Anna Dravida Munnetra Kazhagam (AIADMK), the Marumalarchi Dravida Munnetra Kazhagam (MDMK) and the Desiya Murpokku Dravida Kazhagam (DMDK) opposed the award. However, AIADMK general secretary and former Chief Minister Jayalalithaa, who opposed the award and demanded that the DMK government should resign, wanted the Centre to notify the award in the gazette.

She said the State would lose 20 tmc ft under the final award and attacked the apportionment of 30 tmc ft to Kerala. "Kerala neither is in a position to receive 30 tmc ft nor does it have enough ayacut to irrigate them. This water will go into the sea and be wasted. Tamil Nadu should provide 6 tmc ft of water from the Bhavani river to Kerala. This is great injustice. Farmers fed by Bhavani Sagar will be betrayed in future. All sections of farmers in Tamil Nadu are, therefore, going to suffer," Jayalalithaa said.

MDMK general secretary Vaiko said the final award has come as a shock and a disappointment. "The people of Tamil Nadu have been denied justice." DMDK founder Vijayakant said that while Karnataka had perennial rivers like the Krishna and the Tungabhadra besides the Cauvery, Tamil Nadu had only the Cauvery. "In the final award, we have lost what we had got earlier. This is the truth," Vijayakant claimed.

Karunanidhi and Public Works Minister Durai Murugan refuted Jayalalithaa's claims. Karunanidhi pointed out that if Mettur readings were to be taken into account, Tamil Nadu would get 11 tmc ft more under the final award than under the interim award. "It is funny that Jayalalithaa should demand that the final award should be gazetted even while rejecting it," the Chief Minister said. Durai Murugan argued that Kerala would not receive 6 tmc ft of water from Tamil Nadu's share. "The 6 tmc ft will come from within Kerala's share of 30 tmc ft. Contrary to Jayalalithaa's claims, this water cannot go into the sea. It cannot go back in the same direction from which it comes. This water comes from Kerala to Tamil Nadu. If Kerala does not use it, it will reach Tamil Nadu," said Durai Murugan.

Farmers' organisations such as the Cauvery Protection Committee and the Tamil Nadu Bharatiya Kisan Sangh opposed the award. The former said that a "deeper" reading of the final award would show that it was a setback to the State. The latter feared that the award would render Tamil Nadu a barren State.

Farmers in the tail-end areas of the Cauvery delta were disappointed that the award had not provided for enough water in June when the kuruvai crop is sown.

Section III

Country Experiences

10

Ecology, Planning, and River Management in the United States: Some Historical Reflections

Martin Reuss

River ecologists are also river-basin planners. However, their role in planning has developed slowly over the decades since the beginning of the 20th century. Three major factors explain this phenomenon. First, ecologists focused on plant and animal communities rather than on broader policy issues related to land settlement and water development. Second, the federal government, and most state and local governments as well, used mainly economic criteria to justify projects. Intangible benefits, including the value of species or an aesthetically pleasing landscape, drew relatively little attention. Third, the public generally favored development, especially during the Great Depression of the 1930s. Only after World War II did the public's position shift in favor of more preservation, as ecologists developed the concept of the ecosystem, large dam projects forced basin inhabitants from their homes, and chemical and nuclear pollutants threatened the environment. Also, urbanization increased support for the preservation of

Source: www.ecologyandsociety.org The article was published in Ecology and Society 10 (1):34 © *Martin Reuss. Reprinted with permission.*

recreation sites and of streams undisturbed by human intervention. Meanwhile, partly through important advances in geomorphology and hydrology, ecologists acquired new tools to understand the land-water relationship within river basins. Neverthless, benefit-cost analysis continued to dominate federal water-resources planning, and organizational culture and competing or overlapping bureaucracies hampered rational water resources administration. Environmental groups and physical, natural, and even social scientists began to promote alternative ways to develop rivers. Today, the ideas of integrated water resources management, sustainable development, and comprehensive river-basin management dominate much of the thinking about the future course of river planning in the United States. Any future planning must include ecologists who can help their planning colleagues choose from among rational choices that balance ecological and human demands, provide advice when planning guidance is drafted, assist engineers in designing projects that lead to ecologically responsible solutions, and help monitor results.

Introduction

Ecology is as much, and perhaps more, social science than physical science, because it deals with relationships among all living creatures, including human beings, and their connection with the nonliving world. For ecologists engaged in the restoration of natural systems such as flood plains, the relationship with social science is particularly close. Even the most reductive and mathematical presentations at the Second International Symposium on Riverine Landscapes in Storforsen, Sweden, could not totally ignore issues dealing with social dynamics. Other presentations focused more explicitly on production and consumption and on reconciling human needs with sound ecological science. The common denominator was the implicit insistence that ecological science could help resolve the problems visited upon the world by applied science and engineering. The truth was that nearly all the presenters showed that river restoration was an exercise in planning as much as science, and that the best river restoration ecologists share some of the social science skills and much of the aptitude of professional planners.

This essay attempts to put ecology as a planning tool in some historical context, but first we need to understand the important ways in which ecology departs from many other sciences. Perhaps most significantly, rather than extending mankind's technological control of nature, ecologists often constrain technological applications. Trade-offs are required, and in the process ecologists essentially become nature's advocates. Moreover, ecologists and other scientists may spend a very long time, and often a substantial amount of money, obtaining and accurately measuring objectives, and some results may not even occur in the area being restored. Success often requires adaptive management techniques.

Finally, and most important, ecology, and perhaps especially ecological restoration, threatens the way mankind, at least Western civilization, has historically arranged its affairs. Ecologists do not fear technology but have suggested that technology by itself is rarely the complete answer and may even provoke disaster. Periodic floods and droughts support this view. Added to this technological uncertainty is the question of mankind's role in the divine order. Former US Secretary of the Interior James Watt made reference to this question when he observed that the Bible advises us "to occupy the land until Jesus returns" and that environmentalists are "the greatest threat to the ecology of the West [i.e., of the western United States]" (Reuss 1992). Whether one agrees with Watt or not, ecological restoration may affect not only the geography of the Earth but also the landscape of the mind.

Ecologists on the Margins: Early 20th Century River-Basin Planning

Riverine ecologists deplore the human alterations that have degraded and depleted water, contributed to disease and species loss, increased invasive species, and affected the natural movement of material between ecosystems. Through river restoration and the modification of human behavior, they hope to salvage the advantages of natural systems, such as clean water, species restoration, increased biodiversity, improved public health, and even the economic benefits resulting from a restored commercial fish and wildlife population. These ecologists transcend the boundary between pure science and applied resource management, whether the latter is called "ecosystem management" or something else, i.e., between the science that seeks to understand and the science that seeks to direct. Their rivers are "natural laboratories" to be preserved for science and protected from human disturbances.

This ecological perception goes back a century or so. Barrington Moore, former president of the Ecological Society of America (ESA), which was established in 1915, provided one example. He came before the House of Representatives Agriculture Committee in 1923 to protest the draining of bottomlands along the Upper Mississippi River for agricultural use. He represented 26 different organizations, including the ESA, the Sierra Club, the National Geographic Society, and the American Automobile Association. The ESA, he pointed out, had been working to develop reserves as outdoor laboratories, but drainage activities threatened to destroy the opportunity to learn more about natural processes and functions. Focusing on one stretch of wetlands along the Upper Mississippi, the Winneshiek Bottoms, Moore argued that it had a "remarkable assemblage of plants, fishes, birds, and other animals ... To drain that area, even though it were valuable for agriculture, which it is not, would be the same as destroying a library which contained manuscripts of which there were no other copies" (Anfinson 2003).

Moore's argument probably availed him little in political circles. Former Secretary of the Interior, Franklin K Lane, an attorney by training, exemplified the prevailing climate when he declared in a commencement address at Brown University in 1916, "The mountains are our enemies. We must pierce them and make them serve. The sinful rivers we must curb" (Pisani 2002). For Lane, conservation meant reigning in capricious and wasteful natural forces and managing the environment for the betterment of the human population. His statement seems today both arrogant and subversive, but in fact it mirrored a conservation ethic that typified the Progressive Era (ca. 1900–1920) in the United States and the Age of Positivism in Europe, although no doubt, many would have questioned the ability of rivers to sin. Rather than preserving stretches of river bottomland, "nature's libraries" according to Moore, for study and enjoyment, conservationists urged the maximum exploitation of the nation's freshwater; the less water "wasted," the better.

For many conservationists, the way to reduce "wasted" water was through multipurpose river development. They championed the harnessing of rivers to satisfy a multitude of human needs, including hydropower, navigation, water supply, flood control, and irrigation, although some engineers questioned whether all these needs could be technologically reconciled. For instance, hydropower favors relatively full reservoirs, whereas flood control requires reservoirs to have reserve capacity to handle floods from upstream. In all cases, the emphasis was on production and consumption, i.e., the maximum use of the water resource consistent with its replenishment. The idea grew out of increasing demands for irrigation water in the western United States and for hydropower throughout the country.

Even though multipurpose development clearly subordinated natural processes to human requirements, there is little evidence that ecologists attacked the concept as a whole. Rather, like Moore's argument about the Upper Mississippi wetlands, protests were directed against the destruction of particular regions. Given the dominant ecological focus of the time, this can hardly be surprising. As editor of the journal *Ecology*, Moore had suggested in 1915 that ecology was an integrating science, but his peers generally stayed closer to the ground and limited their studies to individual plant and animal communities. Indeed, another distinguished ecologist, Victor Shelford, in 1919 defined ecology as "the science of communities" (Worster 1977). Highly empirical research in these communities, accurate description, mathematical analysis, and inductive reasoning would lead to new insights about the laws of nature (Kingsland 2004). This approach honored its 19th century forbears, Alexander von Humboldt and Charles Darwin.

The "science of communities" appeared tailor-made to fit regional planning efforts during the New Deal of President Franklin D Roosevelt in the 1930s. New Deal planners believed that regional administration, as exemplified by the Tennessee Valley Authority (TVA), resulted in better integrated and more economical plans. They envisioned flexible administrations capable of responding to social shifts. Often under the umbrella of the Social Science Research Council, numerous social and natural scientists, ecologists among them, became involved. Questions emerged that crossed disciplinary boundaries. Were engineering solutions economically efficient and socially beneficial? How should regions be defined, i.e., culturally, economically, common natural resources, drainage basin, etc.? What should be the objectives of regional planning, and did these objectives threaten traditional institutions and life-styles? Hypotheses about social and administrative behavior abounded, with the TVA serving as a prototype (Reuss 1992).

The TVA itself funded many regional studies. Perhaps this helps explain why so much planning attention is centered on river basins. Other reasons include the ongoing interest in multipurpose river planning and the ability to define a drainage basin unambiguously. Nevertheless, nature's topographic boundaries, no matter how well defined by elevation and water flow, do not necessarily coincide with the boundaries set by technology and economics, not to mention politics. As long as planners focused on navigation and flood control, on the control of the water in short, river basins worked well. However, when they considered hydropower, transportation, agriculture, industry, and land use, the drainage basin worked less satisfactorily, because these activities touched areas that often crossed drainage-basin boundary lines. Geographers in particular became quite enthusiastic about multipurpose river-basin planning, although some political scientists doubted that river basins could serve as "decision arenas" because of congressional resistance and doubts about their economic efficiency (Reuss 1992).

The study of specific plant and animal populations definitely complemented regional planning efforts. Nevertheless, although the data gathered certainly led to a greater understanding of the interdependence within the natural world, ecologists still remained uncertain of the dynamics at work. Some emphasized competition theory in a Darwinian framework; others saw cooperation as the dominant motif even though the tide of fascism and militarism encircling the globe, contradicted such optimism. In short, ecological investigations did not always lead to greater theoretical understanding.

By the late 1930s, however, far more intellectually exciting ideas from the UK attracted growing attention in the United States. In 1927, Cambridge zoologist Charles Elton had published his major work, *Animal Ecology*, which set forth the concept of a food chain linking the smallest to the largest animals. This was an integrated vision of producers and consumers with a place, i.e., a "niche," reserved for every organism in the chain. In this way, Elton defined a species by what it did rather than by its structure. Then, in a 1935 essay, Oxford botanist A G Tansley took Elton's insight a significant step forward, when he dismissed the ideas of both a food chain and separate plant and animal communities. Instead, he proposed the "ecosystem," a concept that integrated living and nonliving substances. Not food but energy flow, i.e., the exchange of energy and of chemical substances such as water and nitrogen, defined an ecosystem. The young American scientist Raymond Lindeman produced a scientific paper in 1942, "The Trophic-Dynamic Aspect of Ecology," that merged Tansley's and Elton's concepts. He suggested that organisms could, in fact, be divided into consumers and producers that are further divided into different levels. Energy is transferred from one level to another, with some energy being lost with each transfer (Worster 1977).

No matter how exciting, the idea of the ecosystem arrived too late in the United States to have any impact on New Deal regional planning studies. Ecology supplied few theoretical tools to the natural resources planner of the 1930s. Ecologists remained marginal players who provided data and interpretation, but not a coherent conceptual framework with which to view river basins. Their impact on river basin planning would come only after the middle of the 20th century.

Science and Scientific Management

In looking back at the development of river basins in the United States, it is easy to exaggerate the influence of the Tennessee Valley Authority (TVA) and other regional planning efforts. The fact is that, until the 1960s, Congress opposed most regional planning efforts and prevented the development of other regional basin commissions based on the TVA model. Neither President Roosevelt nor President Harry S Truman, for instance, prevailed in arguing for a Missouri River Basin Commission. Nevertheless, although Congress opposed political innovation, it embraced with some enthusiasm the idea of scientific management, an idea that had emerged during the Progressive Era. In the aftermath of the catastrophic Dust Bowl of the 1930s, Congress was especially eager to encourage investigations of the relationship between land and water. In 1935, it authorized the creation of a Soil Conservation Service within the Department of Agriculture to help farmers. Then in December 1936, a distinguished group of civil servants, constituted as the Great Plains Committee and chaired by Morris Cooke, head of the Rural Electrification Administration, issued a unanimous report that concluded that the Dust Bowl was caused entirely by misguided agricultural practices reaching back three-quarters of a century. Members embroidered their analysis with a kind of pop ecology: "Nature has established a balance in the Great Plains by what in human terms would be called the method of trial and error. The white man has disturbed this balance; he must restore it or devise a new one of his own" (Worster 1977). If their warning were ignored, the result would be a perennial desert in America's heartland. Scientific management was the key to success.

Government research laboratories took the lead. Gifted scientists and research engineers, some of whom had looked in vain for university positions during the Depression, studied sediment transport, bed movement, and turbulence in rivers. Academic laboratories, some built with New Deal government funding, became allies in this effort. New government flood-control projects put a premium on understanding river behavior. At the Corps of Engineers Waterways Experiment Station in Vicksburg, Mississippi, engineers gained insights into the river meandering process. At a Soil Conservation Service experimental station in Greenville, South Carolina, Hans Albert Einstein, the son of the physicist, and others began to develop a statistical explanation of sediment transport. Across the continent, at another Soil Conservation Service laboratory at the California Institute of Technology, hydrologist Hunter Rouse formulated a groundbreaking equation to describe the distribution of suspended sediment in open-channel flows. All these efforts contributed to a better understanding of the ways in which water and sediment interacted. They all also responded to engineering problems, often resulting from specific project designs or operational challenges, rather than to scientific theory.

At the same time, major advances in geomorphology contributed to the scientific understanding of the drainage basin as an integrated unit. Here, too, the federal government made important contributions. Going back to the 19th century, John Wesley Powell's government-sponsored western explorations had given him the idea that the physical history of a region could be read from a study of its drainage system in relation to its rock and mountain structures (Chorley et al., 1964). Powell's colleague Grove Karl Gilbert saw the Earth's landforms working toward a state of "dynamic equilibrium" in which the forces of erosion, including water, would equal the forces of resistance. Gilbert observed the inverse relationship between amount of water and the slope of the land from a river's mouth to its source. He molded his observation into what he called the "law of divides" and concluded that, in accordance with this law, mountains are steepest at their crests. Indeed, Gilbert maintained, the Earth's topography can be explained solely in terms of the laws he advanced (Gilbert 1877, Chorley *et al.*, 1964).

Gilbert influenced geographer William Morris Davis's concept of a "geographical cycle," an idea that in its emphasis on evolution may have also reflected Darwin's profound impact on late 19th century science. For Davis, the river dominates the landscape, passing through successive states of youth, maturity, and old age. He spoke of a "geographical cycle" that melded geological structure and erosive processes into an ever-changing landscape. If you can identify the landscape, you will know its age (Chorley *et al.*, 1964). Gilbert and Davis provided descriptive and theoretical tools that profoundly changed hydrology. Equally important, in linking land and water, they compelled hydrologists to appreciate the entire river basin, not just the water running through it. Their influence was so pervasive and their writing so compelling that their descriptive approach held sway in hydrology for more than 50 years.

Only in the mid-20th century did engineers, hydrologists, and geomorphologists successfully introduce quantitative analysis into geomorphology. This transformation produced a much more useful tool for understanding the development of drainage basins and stimulated important conceptual advances in both hydrology and ecology. Indeed, it helped render hydrology as an acceptable science in the classical sense by illuminating fundamental natural laws and relationships, rather than simply functioning as an adjunct of engineering, more narrowly focused on the accumulation, interpretation, and application of data for water projects. Hydrologist Robert Horton initiated this transformation in 1945, when he published a 95-page article with 40 figures in which he proposed a new "law of stream lengths" that showed a constant relationship between the number of streams of different orders, from main stem to subtributaries, within a basin (Horton 1945).

Horton's article initiated a general reevaluation of Davis's historical and qualitative geomorphology. The influential American geomorphologist Arthur N Strahler decried Davis's methodology as "a superficial cultural pursuit of geographers which is completely inadequate as a natural science" in a rather courageous speech in 1950 before the Association of American Geographers (Strahler 1950). He later observed that "Davis himself maintained that the aims of his method were geographic; that the consideration of process was introduced merely to permit an orderly genetic system of landform classification." However, if geomorphology were ever to achieve full stature as a branch of geology, "it must turn to the physical and engineering sciences and mathematics for the vitality which it now lacks." According to Strahler, the successful geomorphologist will be "trained as a geologist, (who) has built up a life-time store of information and experience, much of it relating to theoretical and historical aspects of geology." Geomorphologists, Strahler proposed, one should study processes and landforms "as various kinds of responses to gravitational and molecular shear stresses," develop quantitative determinations of landform characteristics and causative factors, formulate empirical equations using mathematical statistics, build on the concept of open dynamic systems, and, finally, "deduce general mathematical models to serve as quantitative natural laws" (Strahler 1952, Sack, 1992).

Luna Leopold carried Strahler's battle to the hallways of the US Geological Survey, where he began work in 1950 just as he was completing his doctorate at Harvard. He soon found himself out in Wyoming with his friend, John Miller, another Harvard Ph.D. Together, decades before such investigations became common, they studied the impact of climate change on river valleys. Along the way, Leopold became intrigued with the question of why rivers have particular widths. He and another colleague, Thomas Maddock, Jr., completed a paper on "The Hydraulic Geometry of Stream Channels and Some Physiographic Implications" in 1953 that was published by the Geological Survey. In this paper, Leopold and Maddock showed a mathematical relationship between river width and discharge. The paper initiated a new approach to fluvial geomorphology called "hydraulic geometry." In 1964, the book *Fluvial Processes in Geomorphology* was published, co-authored by Leopold, M Gordon Wolman, and Miller, who died before the book went to press. The book quickly established itself as a basic work on the subject, and Leopold's integration of geomorphology into hydrology helped put hydrology on a sound scientific footing within the federal government (Leopold and Maddock, Jr 1953, Leopold *et al.*, 1964).

By the middle of the 1950s, new and exciting concepts and research had energized the ecological community. Developments in geomorphology and hydrology contributed a much more sophisticated understanding of the natural operation of the drainage basin, and the idea of the "ecosystem" provided a holistic explanation of the interdependency of living and nonliving substances. However, unlike economists and geographers, ecologists did not appear eager to translate new concepts and theories into government policy. The nation's stewardship of its natural resources deferred more to economics, engineering, and politics than to science.

Benefit-Cost Analysis and Multi-Objective Planning

In 1933, Luna Leopold's father, Aldo, wrote an essay on "The Conservation Ethic" that criticized those who valued land strictly in economic terms. Subsequently, the renowned forester and game management expert, and later president of the Ecological Society of America, increasingly viewed scientific management with suspicion because of its emphasis on economic rather than ecological benefits (Worster 1977). He also implicitly insisted that mankind was part of nature and not outside of it. At the time, even some of Leopold's ecological colleagues casually wrote of human impacts on ecosystems as though natural ecosystems excluded humans; conversely, any human intervention rendered ecosystems artificial. In the two decades following World War II, Rachel Carson's *Silent Spring* (1962), fears of nuclear devastation, and real and imagined chemical threats to land and water strengthened this perception. It appeared that humans were incapable of improving the natural world; they could only degrade or destroy it. Indeed, some would go further and argue that any change to the natural world degraded both the world and its corrupters. That position influenced both ethics and science, and only during the environmental era that began around 1970 did that perception change. Today, most ecologists accept humans as an integral part of ecosystems, for good or ill.

Nevertheless, economics, not ethics, has justified water projects in the United States since the origins of the Republic. As Leopold observed, nature had become commodified, and the value of flora and fauna, water and land, was reduced to dollars. Congress completely embraced the concept in the Flood Control Act of 1936, which for the first time stipulated that flood control "was a proper activity of the Federal Government in cooperation with States, their political subdivisions, and localities thereof." The Act required that no flood control project should be built if the benefits "to whomsoever they may accrue" did not exceed costs. Although born of a need to rationalize the planning process in some way, benefit-cost analysis proved devilishly hard to implement in practice. The planning community faced fundamental problems in developing rational and equitable procedures. Questions ranged from definitions of national interest or flood "damage" to what constituted tangible and intangible benefits. Intangible benefits might include the value of a deer, woodlot, or wetland; the value of recreation or of a positive aesthetic experience; or even the value of a human life. In all cases, value was to be transformed into market price. Subsequent amendments and modifications to the Flood Control Act eventually extended benefit-cost analysis to all federal water resources projects, including large multipurpose dams (Arnold 1988, Reuss 1991, 1992).

Benefit-cost analysis essentially marries market principles to the utilitarian philosophy of Jeremy Bentham. It suggests that obtaining "the greatest happiness for the greatest number" is a matter of using available resources to optimize social welfare benefits. In the 1930s, a popular economic idea was "Pareto Efficiency," in which no individual could be made better off without leaving someone else worse off, but it was basically naïve, because few policy changes involve situations in which literally no one loses. The major advantage of benefit-cost analysis is that it imposes discipline on public choice so that scarce resources are rationally allocated in ways that ensure their highest possible value. It also provides a means for addressing social problems systematically. In fact, benefit-cost analysis in the abstract negates the necessity for any public input.

The problem is that, in democratic countries, public interests and pressure groups distort this process, for better or worse. The economists' rule of reason is transformed into the rule of consent. The rule of reason is further imperiled when normative beliefs impinge on a process that is supposedly non-normative, i.e., based on market price. Any decision on the value of human life or of a particular habitat, for instance, must be based on some normative calculation. One can go still further and argue that non-normative calculations based on actuarial standards and economic data will still result in normative solutions. For example, the use of historical data on farm production or the wealth of different populations would tend to ensure that the future would replicate the past; higher net project benefits would be assigned to higher net worth, whether of land or people. None of these methodological obstacles would in themselves doom benefit-cost analysis, if the calculations were used only as a guide and framework for addressing social concerns. The methodology might then be considered simply a pragmatic and even morally responsible approach. Instead, benefit-cost analysis became the basis for establishing formal federal decision-making criteria. Against a background of changing values, new technology, and new science, experts and policy makers hammered out these criteria over the two or three decades following passage of the Flood Control Act of 1936 (Byrne 1987). Despite changes in the criteria to embrace, or at least acknowledge, emerging environmental values, benefit-cost analysis, with its heavy emphasis on market price, became a red flag teasing a confused and angry public.

As practiced, benefit-cost analysis raised substantial ethical issues almost from the time the Flood Control Act enshrined the practice. Economists insisted that the social value of income generated by a project remained the same whether the project benefited poor or wealthy people. The ethics of such a position even bothered some of its proponents. Do, for instance, benefits "to whomsoever they may accrue" require that the protection of a millionaire's property on one side of the river be considered more beneficial than saving 10 poor hovels worth far less money on the opposite bank? This may be sound economics, but it is usually bad public policy. Benefit-cost calculations also require that social problems be treated independently to derive their "cost," when more often than not they are interdependent, e.g., disease, poor quality water, and inadequate education. Also, alternative solutions to social problems need to be commensurable and finite to make comparisons, a stricture that, among other things, leads to a bias in favor of structural vs. nonstructural, e.g., restrictions on land and water use and on types and number of buildings, and flood-control solutions.

Nevertheless, probably the most fundamental objection to benefit-cost calculation is the inappropriate application of market prices. Other considerations aside, benefit-cost analysts would inevitably conclude that rich farmland should be protected before poor agricultural land or that urban areas should receive pollution protection before rural areas. However, public input and interest groups can change the calculus. Implicit recognition is given to the fact that social problems and objectives, including those associated with river control, are not only interdependent, but at least partially subjective. Given this, benefit-cost analysis, i.e., the rule of reason, is a poor decision-making tool. It often favors the *status quo,* fails to answer questions about equity and environmental justice, and assumes an economic climate at variance with reality.

A group of academicians at Harvard University thought that the innovative use of a new machine, the electronic computer, might resolve some of these problems. In 1955, they formed the Harvard Water Program, a multidisciplinary research and training program to develop new methodological techniques for water resources planning. The professors and their students used computers to develop physical and economic simulations of water systems. They also developed something that eventually was called "synthetic hydrology" – computer-generated predictions of future hydrologic activity and impacts. This approach reduced reliance on historical data and integrated physical, social, and economic data into the programs. Finally, they developed a new approach to economic evaluation known as "multiobjective analysis."

Unlike benefit-cost analysis, which always seeks economic efficiency, multiobjective analysis designs water systems to address all the objectives sought by planners, including non economic values such as environmental quality or even the preservation of an ethnic neighborhood. While relying on computer simulations to identify the consequences of various options, it concedes that inevitably the decisions must be left up to the politicians or planners. Concerned groups must sit down, discuss the options, and determine trade-offs. This approach forces the various interests to develop priorities and negotiating positions. It demonstrates the need for interdisciplinary and intergovernmental cooperation in water resources planning (Reuss 1992).

However, this approach was expensive. Canvassing a broad range of options for the development of an entire river basin required considerable time and resources and often resulted in plans for projects that Congress refused either to authorize or to fund. In the late 1960s and early 1970s, the Bureau of the Budget, which after 1970 was known as the Office of Management and Budget, believed that multiobjective "framework" studies were grossly wasteful and refused to support them. Arguments that the studies would lead to better projects and more equitable solutions did not prevail. Given this situation, Congress, not the bureaucracy or technical experts, remained the great arbitrator. After 150 years of water resources development and a hodgepodge of statutes and executive orders, the United States still had no institutional framework for developing comprehensive water resources programs.

Water Resources Planning in the Environmental Era

Today, water planners in the United States and around the world are attempting to develop "comprehensive" water management plans. These plans are meant to be applied to the entire watershed and are sometimes described as exercises in integrated water resources management (IWRM). The idea is to develop projects that contribute to economic development while respecting ecosystems and not degrading the environment for succeeding generations. This definition also applies to the concept of "sustainable development." Unquestionably, the two ideas have become linked in the public mind. Nevertheless, the conceptual allure of sustainable development has tarnished over time as it has become apparent that the term means different things to different people. Indeed, it can be manipulated in ways that may contribute more to political than ecological stability. Discussion often centers on the degree of allowable environmental degradation rather than the extent of necessary conservation practices.

Another imprecise term is "ecosystem management," which was popularized by Eugene P and Howard T Odum in the 1960s and early 1970s. In part, their work reflected the influence of their father, distinguished sociologist Howard W Odum, and his groundbreaking studies on regionalism. Whereas the social science concept of regionalism recognized and attempted to preserve both cultural and natural regional characteristics, the Odums argued that ecosystem management holistically integrated physical and biological elements in the natural environment (Odum 1975). "Sustainability" is an objective of ecosystem management, just as it is for IWRM. Difficult questions remain concerning the priorities that should be given to various ecosystem functions and the degree to which they can actually be used to administer land and water.

Ecosystem restoration compounds the problems facing water resource planners. Restoration calls for a definition of a "natural system," but the term's imprecision leads to political consequences, matching agency against agency and interest against interest. Moreover, as science historian Sharon Kingsland has observed, it is not all that easy to reconstruct a significantly altered system. She notes that "nature will not automatically 'bounce back' and return to its former self" (Kingsland 2004). She might have added that it is not always possible to know what the "former self" was. Given the normal evolution of rivers basins, the proper term may, in fact, be "former selves." Is a "natural system" one that predates human intervention, one that has reached some sort of ecological balance, or something in between? In other words, what baseline data should be used? The answer could spell the difference between a $10 million or a $100 million project. Are stable systems composed of diverse species less vulnerable to outside interference than simpler systems or systems still not in balance? Finally, what assumptions should be made about mankind's future relationship with land and water, and what role will new technology or even new political arrangements play? Engineers occupy the central role in the design of water projects, but ecological input is critical in decisions regarding the type and degree of intervention to restore a system.

The challenges facing ecologists should not be understated. Despite the strides in integrating ecological concepts into river basin planning, the reality is that the technical literature inevitably favors hard engineering data. This may be especially true of flood control projects. In justifying such projects, traditional planners focus on potential flood damages measured in dollars and cents, whereas the scientific community addresses vague issues that are long-term and often require subjective evaluation. Additionally, scientific investigations are often expensive and time-consuming; meanwhile, the flood threat remains. In short, engineers and economists use historical data and empirical analysis to predict the socioeconomic consequences of the "without project condition." They confidently talk of the facts of flooding, i.e., the potential damages to life and property, while scientists argue about values, some of which cannot easily be empirically verified. Engineers remain cautious technological optimists who identify attainable goals, whereas scientists often only imperfectly define objectives.

In the United States, the bureaucratic response to water issues depends very much on organizational culture. In the period from about 1965 to 1985, a kind of institutional dissonance existed. With a tradition firmly linked to economic development, engineering agencies at the state and federal levels lagged in responding to emerging public environmental values. Perhaps this lag was predictable. Agencies not only have missions; they have attitudes. The Corps of Engineers, which arguably changed faster than most development agencies within the federal government (Mazmanian and Nienaber 1979), nevertheless used numerous civilian engineers and Army officers who believed that the agency's mission was to "improve" rivers for navigation, flood control, and hydropower, not to compromise engineering safety or economic efficiency for the sake of preserving ecological services. Eventually, their attitude changed, and today the Corps of Engineers counts the "environmental mission" among its primary missions.

Although progress has been made in reorienting organizational cultures, efforts to eliminate organizational dysfunction such as the bureaucratic barriers that impede rational, scientifically valid administration have not been equally successful. For example, the hydrological cycle describes the integrated, interdependent circulation of water under, on, and above the Earth. Groundwater, surface water, and precipitation combine to determine river flows, floods, and droughts. Nevertheless, no one agency oversees all aspects of the water cycle. Instead, in the United States numerous agencies are involved. At the federal level alone, these include the Weather Service, the Geological Survey, the Environmental Protection Agency, the Army Corps of Engineers, the Natural Resources Conservation Service, and the Bureau of Reclamation. Literally, dozens of agencies and congressional committees and tens of thousands of people are involved in determining the regulation of river flows, pollution standards, flood insurance standards, and water availability for numerous purposes. No efficient, integrated approach to water control and development can come from such a large, overlapping, bureaucratic structure, and none seems likely in the foreseeable future.

Conclusion: A Look Forward

Competition and cooperation are not only present in ecosystems, but they also define bureaucratic organizations. In the case of the natural resource agencies charged to protect and/or develop the environment, the success or failure of ecosystem management might directly affect the stability of both the environment and the agencies themselves. Ecological stability might lead to greater cooperation among state and federal resource agencies. Conversely, ecological instability could threaten bureaucratic agreements and entice agencies to resurrect earlier, competing plans that reflect particular missions and organizational cultures. Planners and politicians may raise new arguments about appropriate flow lines at different times of the year or the merit of various ecological values. Old questions about how much a tree, flower, fish, or deer is worth will once again attract attention, no doubt with little consensus as a result. Attempts to remedy damage to an ecosystem will raise objections among some ecologists who oppose any mitigation that falls short of restoring the "natural system." In fact, the mitigation may actually introduce new problems.

In effect, science has changed the relationship between the agencies and their missions. Greater ecological awareness forces both an ethical and professional transformation. The ethical change is undoubtedly the more difficult challenge. It requires the reconciliation of two seemingly disparate objectives: defense of the environment and the satisfaction of a broad range of human needs. Neither blasé insistence, that the objectives are compatible nor the philosophical argument that nature has rights will likely force change in mankind's attitude toward the environment. Instead, an environmental ethic must evolve that reconciles public choice, economic efficiency, and some sort of moral intuition. Balance is necessary, and in a democracy balance is always a moving target. Projects need to be affordable and supported by a broad range of diverse interests. Properly understood, an environmental ethic will reduce demands for water, stress that water is a finite resource, and insist that competing interests acknowledge mutual obligations for the sake of regional harmony. Constraints must be placed on public choice.

This approach to water resources planning can cause substantial turmoil within a water development agency. It requires new kinds of expertise and places more emphasis on process and less on product. Successful projects are not simply economically efficient but respond to a broad range of objectives with the least disturbance to the environment. Such projects may be non-structural in nature and may even be as simple as new water-use agreements among regional parties. Planning is always holistic and comprehensive. Given this approach, ecologists must help transform agencies without sacrificing agency morale and technical expertise. They can help their planning colleagues choose from among rational choices that balance ecological and human demands, provide advice when planning guidance is drafted, assist engineers in designing projects that lead to ecologically responsible solutions, and help monitor results. For a truly sustainable world, ecologists must be firmly integrated into the planning process.

Acknowledgments

The views expressed in this article are those of the author and do not necessarily reflect those of the US Army Corps of Engineers, the Department of the Army, or the Department of Defense.

(Martin Reuss is the senior civil works historian in the Office of History, Headquarters, US Army Corps of Engineers, where he specializes in the history of flood control, navigation, and hydraulic engineering.)

Literature Cited

Anfinson J 2003, *The river we have wrought: A history of the Upper Mississippi*, University of Minnesota Press, Minneapolis, Minnesota, USA.

Arnold, J L 1988, *The evolution of the 1936 Flood Control Act*, Office of History, U S Army Corps of Engineers, Fort Belvoir, Virginia, USA.

Byrne J 1987, Policy science and the administrative state: The political economy of cost-benefit analysis, Pages 70-93, in F Fischer and J Forester, editors. *Confronting values in policy analysis: The politics of criteria.* Sage yearbooks in politics and public policy, Volume 14, Sage Publications, Newbury Park, California, USA.

Carson R 1962, *Silent spring*, Houghton-Mifflin, Boston, Massachusetts, USA.

Chorley R J, A J Dunn, and R P Beckinsale, 1964, *Geomorphology before Davis: The history of the study of landforms or the development of geomorphology, Volume 1*, Methuen, London, UK.

Elton C 1927, *Animal ecology*, Macmillan, New York, New York, USA.

Gilbert G K 1877, *Report on the geology of the Henry Mountains*, US Department of the Interior, US Geographical and Geological Survey of the Rocky Mountain Region, Washington, D C, USA.

Horton R E 1945, Erosional development of streams and their drainage basins: hydrophysical approach to quantitative morphology, *Bulletin of the Geological Society of America* 56:275-370.

Kingsland S 2004, Conveying the intellectual challenge of ecology: an historical perspective, *Frontiers in Ecology and the Environment* 2:367-374.

Leopold A 1933, The conservation ethic, *Journal of Forestry*, 31:634-643.

Leopold L B and T T Maddock Jr. 1953, *The hydraulic geometry of stream channels and some physiographic implications*, Professional paper 252. United States Geological Survey, Washington, D C, USA.

Leopold L B, M G Wolman, and J P Miller, 1964, *Fluvial processes in geomorphology*, W H Freeman, San Francisco, California, USA.

Lindeman R 1942, The trophic-dynamic aspect of ecology, *Ecology* 23:399-417.

Mazmanian D A, and J Nienaber 1979, *Can organizations change: Environmental protection, citizen participation, and the Corps of Engineers*, Brookings Institution, Washington, D C, USA.

Odum E 1975, *Ecology: The link between the natural and social sciences*, Second edition, Holt, Rinehart and Winston, New York, New York, USA.

Pisani D J 2002, *Water and American government: The Reclamation Bureau, national water policy, and the West, 1902-1935*, University of California Press, Berkeley, California, USA.

Reuss M 1991, *Reshaping national water politics: the emergence of the Water Resources Development Act of 1986*, Institute for Water Resources, US Army Corps of Engineers, Alexandria, Virginia, USA.

Reuss M 1992, Coping with uncertainty: Social scientists, engineers, and federal water resources planning. *Natural Resources Journal* 32:101-135.

Sack D 1992, New wine in old bottles: The historiography of a paradigm change. *Geomorphology* 5:251-263.

Strahler A N 1950, Davis' concepts of slope development viewed in the light of recent quantitative investigations, *Annals of the Association of American Geographers* 40:209-213.

Strahler A N 1952, Dynamic basis of geomorphology, *Bulletin of the Geological Society of America* 63:923-938.

Tansley A G 1935, The use and abuse of vegetational concepts and terms, *Ecology* 16:284-307.

Worster D 1977, *Nature's economy: a history of ecological ideas*, Cambridge University Press, Cambridge, UK.

11

The Three Gorges Project and Flood Control of the Yangtze River

Qihui Yu, Yongjun Zhang and Weiyu Tang

The most important task of the regulation and development of the Yangtze River Basin is flood control. The plains in the middle and lower reaching of the Yangtze river are the key areas of flood control. The Three Gorges Project (TGP) is vitally important and backbone project in the development and harnessing of the Yangtze river. TGP is a multi-purpose hydro-development project producing comprehensive benefits mainly in flood control, power generation and navigation improvement. The paper mainly analyses the role of the TGP in the flood control of the Yangtze river.

1. Introduction

1.1 Introduction of the Yangtze River

The Yangtze river is the longest river in China and the third-longest in the world. It originates from the Tuotuo river on the south-western side of the snow-draped Geladandong—the main peak of Tanggula Mountains of the Qinghai and Tibet plateau; it flows through provinces or autonomous regions of Qinghai, Tibet,

---

Sichuan, Yunnan, Chongqing, Hubei, Hunan, Jiangxi, Anhui and Jiangsu, and finally enters into the East China Sea in Shanghai. It has a length of 6,300 kilometres and a catchment area of 1.8 million square kilometres, which is equivalent to 20% of the total land of China. The river has over 700 tributaries and the main tributaries are eight rivers of the Yalongjiang, the Minjiang, the Jialingjiang, the Wujiang, the Xiangjiang, the Yuanjiang, the Hanjiang and the Ganjiang – all have a catchment area of above 80,000 square kilometres. The upper stream of the river is from the source to Yichang with a length of 4,510 km and a catchment area of 1 million square kilometres. The famous "Three Gorges" of 209 km long is just located at the end of this section. From Yichang to the Poyang lake, the river enters into the middle reaches with a length of 940 km and a catchment area of 0.68 millon square kilometres. Down from the mouth of Poyang lake is the lower reaches of the river with a length of 850 km and a catchment area of 0.12 millon square kilometres. From Zhicheng in Hubei Province to Chenglingji – outlet of the Dongting lake – the river has a high degree of risk of flooding in this section because lowlands are scattered along the river. In the middle and lower reaches of the river, many big lakes are connected, namely the Dongting lake, the Poyang lake, the Cao lake and the Tai lake.

1.2 Introduction of the Three Gorges Project

The TGP is a vitally important and backbone project in the development and harnessing of the Yangtze river. TGP is the largest water conservancy project ever built in China, and so in the world. TGP is a multi-purpose hydro-development project producing comprehensive benefits mainly in flood control, power generation and navigation improvement.

The dam site is situated at Sandouping of Yichang City, Hubei Province, about 40 km upstream from the existing Gezhouba Project.

The project is composed of a concrete gravity dam, two power plants and navigation facilities. The dam is of a concrete gravity type. The total length of the dam axis is 2,335 m, with the crest elevation at 185 m and a maximum height of 181 m. The spillway dam, which is located in the middle of the river channel, is 483 m long in total. On both sides of the spillway dam section, there are intake-dam and non-overflow dam sections.

Two powerhouses are placed at the toe of the dam, one on each side. There are 26 sets of turbine generator units in total, 700 MW each, totalling 18,200 MW in installed capacity.

The permanent navigation structures consist of a permanent ship lock and a ship lift. The ship lock is schemed as a double-way five-step flight locks, each lock chamber is dimensioned at 280×34 ×5 m (i.e., length × width ×water depth) capable of passing 10,000 tons of barge fleet. The ship lift is designed as a one-step vertical hoisting type with a container sized 120×18×3.5m, capable of carrying one 3,000 tons passenger or cargo boat each time.

With the Normal Pool Level (NPL) at 175 m, the total storage capacity of the reservoir is 39.3 billion m^3, in which 22.15 billion m^3 is flood control storage.

2. Key Areas of Flood Control in the Yangtze River Valley

2.1 The Integrated Development Plan of the Yangtze River Valley

The "Brief Report of The Integrated Development Plan of the Yangtze River valley"(Brief Report), revised in 1990, was principally ratified by the State Council. The main parts of the plan are:

- To continuously improve the flood control capability of the Yangtze river, and mitigate and eliminate flood disasters.
- To vigorously tap hydropower resources, and boost the integrated usage of water resources.
- To sufficiently use the potential navigation advantages of the Yangtze river and its branches.
- To continually develop irrigation, and enhance water and soil conservancy.
- To realize the water transportation from the south to the north between river valleys.
- To protect water resources, guarantee the water usage of cities and industries, plan the layout of cities and towns along the river, and develop fishery and tourism.
- One can see that flood control is the first task of the regulation of the Yangtze river.

2.2 Key Areas of Flood Control in the Yangtze River Valley

The plain area in the middle and lower reaches of the Yangtze river, filled up by the silt carried by the main river and its branches, is the key area of flood control. It is flat, fertile and sufficient in water resources. Through the long history, 5.53 million hm2 arable land has been cultivated. Now in this area there are 103 million people, lots of cities and towns, factories and industries, and transport facilities. It is one of the most populous and developed areas in China, and has a very important status. Because the land elevations are lower than the flood water levels for 5~6m, even 10m, and because the influx of the flood from the main river and its branches, the plain area in the middle and lower reaches of the Yangtze river shares the most frequent and disastrous flood. Consequently, this area becomes the focus of flood control in the whole Yangtze river valley.

3. Floods and Flood Disasters in Middle and Lower Yangtze River Valley

3.1 Formation and Type of the Yangtze River floods

On the whole it is rainstorms that form the floods in the Yangtze river valley. Except in the Qinghai and Tibet plateau, rainstorms occur everywhere in the valley. The time and regional distributions of the floods in the Yangtze river valley are consistent with those of the rainstorms. Rainstorms in the Yangtze river valley concentrate mainly from May to October. Usually the rainy season of the middle and lower reaches is earlier than that of upper reach, and the south – earlier than the north. In normal years the flood peaks of tributaries do not encounter each other, and the main river in the middle and lower reaches can hold the floods from upper reaches and from the branches in the middle and lower reaches of the Yangtze river, so big flood will not occur. But if the meteorological conditions are not normal and floods from the main river and its branches will occur, this threatens the plain area in the middle and lower reaches of the Yangtze river.

According to the regional distributions and covering areas of rainstorms, usually the Yangtze river floods can be categorized into two types. The first type is regional flood, which is caused by the rainstorms in the upper reach or by the very intensive rainstorms of the middle reach branches such as Hanjiang river and Lishui river.

Historical floods in 1860, 1870, 1935, 1981 and 1991 can be categorized into this type. The other type is whole valley flood. In some tributaries the rainy seasons come earlier or later, and rainy seasons of the upper, middle and lower reaches encounter each other, which causes floods, with high flow discharge and long duration. The floods in 1788, 1848, 1849, 1931, 1954 and 1998 can be categorized into this type. The middle and lower reaches have to hold the floods from upper reach and from the branches of the middle and lower reaches. Any types of flood impose very serious threats on the plain area in the middle and lower reaches of the Yangtze river.

3.2 Flood Characteristic in the Middle and Lower Reaches of the Yangtze River

The floods of the middle and lower reaches of the Yangtze river mainly come from the upper regions of Yichang. The floods from the Dongting lake valley and Poyang lake valley are its important parts, and the flood from the Hanjiang River is also its key sources. Usually the flood upper of Yichang is about 50% of the total flow volume of the Yangtze river, about 90% of the Jingjiang river reaches. The Jingjiang river reaches include parts of the key area of flood protection. Hence upper reach flood of the Yangtze river is most important cause to the disasters of the area in the middle and lower reaches. The flood volumes in Dongting lake valley and Boyang lake valley are also very big, and the flood disasters in these two regions are also very serious.

3.3 The Flood Disasters in the Middle and Lower Reaches of the Yangtze River

The years of 1931, 1935, 1954 and 1998 saw heavy floods leading to severe floods in the middle and lower reaches of the Yangtze river. All these floods caused serious losses.

The flood in 1931 covered 186 counties and cities, which are located in Hubei, Hunan, Jiangxi, Anhui and Jiangsu. It inundated farmlands of 3.39 million hm^2. Around 28.5 million people were affected by the flood, and 145 thousand lives were claimed. It was Hubei province and Hunan province that shared the heaviest losses. In 1935, the five provinces mentioned above encountered floods again. 1.51 million hm^2 farmlands were flooded, 11.03 million people were affected and 140 thousand lives were claimed. Again, the mostly affected provinces were Hubei province and Hunan province. The floods in 1954 covered 123 counties and cities, which are situated in provinces like Hubei, Hunan, Jiangxi and Jiangsu. 3.17 million hm^2 farmlands were flooded and 18.88 million people were affected and 30 thousand people died. The railway track from Beijing to Guangzhou could not be restored for about 100 days. The flood disaster areas in 1998 covered 334 counties and cities, inundated farmlands of 240 thousand hm^2 and affected 2.316 million people. Among the affected provinces, Hunan, Hubei, and Jiangxi shared the most serious disasters.

4. Standards for Flood Control of the Middle and Lower Reaches of the Yangtze River

The main stream of the middle and lower reaches of the Yangtze river receives water from the upper reaches and from tributaries in the middle and lower reaches of the river. This results in different flood resources and complex components of floods. For a single flood, the return periods at different stations are usually quite different, and at the same station the peak flow discharge and flow volume in different duration also share very different return periods.

This can be illustrated by two big floods that occurred in the 20th century.

- In 1954, the peak flow discharge at Yichang Station was 66,800m^3/s with a return period of only about 10 years, while the return period of the flow volume for 7 days was about 30 years, for 15days was close to 85 years and for 30 days with a flow volume of 138.6 billion m^3 was about 100 years. At Luoshan station and Hankou station the return period of the flow volume for 30 days was approximately 200 years.

- In 1998, the peak flow discharge was 6,3300m^3/s at Yichang Station with a reoccurrence of 6~8 years, and the flow volume of 30 days with 137.9 billion m^3 had a return period of nearly 100 years, while the return period of the flow volume of 30 days at Hankou Station was only approximately 30 years.

Hence, one could not coarsely define the reoccurrence of the flood that occurred in the past or will occur in the future. There are very big variations among different stations or hydrological factors. Even for stations at the mainstream, for instance, the return period of the flood in 1954 varied from 10 years to 200 years, and in 1998 varied from 6 to 8 years to 100 years. The floods of a certain frequency (reoccurrence) should not be fixed as a certain mode. Therefore, for flood control of the area of the middle and lower reaches of the Yangtze river, it is difficult to define a general design flood with normal concepts, which characterize as a certain return period. After years of study, the flood in 1954, the biggest one in the 20th century, is taken as the general standard for flood control of the middle and lower reaches of the Yangtze river.

The flood control of the Jingjiang river section is governed by the peak flow discharge of upper reaches of the Yangtze river. Because the upper reach floods are almost not impacted by human activity such as diversion, dike breaches and regulation, etc., the Jingjiang river section can use frequency for calculation of floods. In 1954, the peak flow discharge in Yichang was 66,800m³/s with a return period of about 10 years, and in Zhicheng, the control point for flood protection was 71,900m³/s with a return period of about 15 years. Obviously it is too low to use these two standards for flood control.

Considering the importance of the Jingjiang river section and seriousness of flood disasters, the standard for flood control is proposed to be 100 years, that is, to defend against peak flow discharge in Zhicheng with a return period. At the same time considerations should be given to the two special big floods, which occurred in 1860 and in 1870, respectively. Both the peak flow discharges in Zhicheng were about 11000m³/s. Once such floods occur again, the area of Jingjiang river section will face grievous disasters, and even the whole area of the middle reaches of the Yangtze river will encounter extremely serious results. So, on the basis of meeting 100-year reoccurrence, when encountering floods like 1870, Jingjiang river must have reliable ability to prevent both dikes from overflowing and breaching, which leads to grievous disasters.

5. Flood Control System of the Middle and Lower Reaches of the Yangtze River and its Current Capacity

5.1 Flood Control System of the Middle and Lower Reaches of the Yangtze River

The "Brief Report" proposed that "applying both detention and discharge and giving priority to discharge" is the guideline for flood control of the middle and lower reaches of the Yangtze river, and consideration should also be given to the principles of benefiting both the Yangtze river and lake along the river, of benefiting both sides of the river, and of coordination among the upper, the middle and the lower reaches. By use of structure measures and non-structure measures, a relatively good solution will be got for the flood problems in the area of the middle and lower Yangtze river. The structure measures include reasonable reinforcement and heightening of the embankments, river course improvement, arrangement and construction of diversion and detention areas and gradual construction of reservoirs on the river and its branches. Among the structure measures, the Three Gorges Project is the backbone and the embankments are the basis.

Under guidance of the plan, projects of improving and strengthening the embankments in the Yangtze river Basin have been launched, focusing on the middle and lower reaches; and some flood diversion and detention projects have been arranged and built by using lakes and depressions along the middle and lower reaches. Additionally, with the economic development, combined with development of irrigation systems, some comprehensive reservoirs with the function of flood control have been built. So far, the total length of embankments along the middle and lower reaches of the Yangtze river, including the Yangtze embankments, embankments along the main tributaries, as well as embankments around Dongting Lake and Poyang Lake, amount to about 30,000 km; and there are 40 flood diversion and detention areas with a capacity of 50 billion m^3 along the middle and lower reaches of Yangtze river, of which, four are in the Jingjiang area, 25 in areas near Chenglingji, six in areas near Wuhan and five in areas near Hukou; 48,000 large-, middle- and small-scale reservoirs, with a total storage capacity of about 160 billion m^3, have been built on tributaries of the Yangtze river. Among these reservoirs only Danjiangkou reservoir on the Hanjiang river and Jiangya reservoir on the Lishui river have the flood control as there first tasks. Flood control system in the plain area of the middle and lower reaches of the

Yangtze river, based on the embankments, with some capacity of flood control, have been primarily formed. It has withstood big floods since 1954 and brought great economic benefits.

5.2 Current Capacities of Flood Control

After years of construction, the current capacities of the flood control of the main river and its branches are as following. Depending on the embankments, the Jingjiang river section can defend against floods of 10-year reoccurrence. By use of flood diversion and detention areas, the Jingjiang river section can defend floods with a return period of 40 years. With embankments, the river section of Chenglingji can defend against floods with a return period of 10-15 years. By relatively optimal use of flood diversion and detention areas, it can meet the demands to control floods of the type of the year 1954. With embankments the river section of Wuhan can defend against floods with a return period of 20-30 years. After the relatively optimal use the flood diversion and detention areas along and before this river section, it can meet the requirements to control floods of the type of the year 1954, whose maximum flow volume of 30 days has return period of 200 years. With embankments the river section of Hukou can defend against flood with a return period of 20 years. After the relatively optimal use of the flood diversion and detention areas before and along this river section, it can meet the requirements to control floods of the type of the year 1954.

6. Flood Control Planning of the TGP

According to the plan of integral flood control of the middle and lower reaches of the Yangtze River, the flood control plan of TGP are:

- With the regulation of the TGP, to ensure a 100-year flood control standard for the area of Jingjiang River;
- Once occurrence of floods of 100-year reoccurrence or floods like 1870, to control the flow discharge at Zhicheng not bigger than 80000m3/s;
- With the co-ordinated operation of the planned flood diversion and detention areas, to guarantee the flood safety of Jingjiang River section and avoid grievous disasters like dike breaches of the Jingjiang River;

- Provided the achievement of the aims mentioned above, it is better to obtain a big reduction of diversion flow near the area of Chenglingji.

According to the requirements of flood control of the middle and lower reaches of the river, two modes of compensation regulations are proposed for flood control operation of TGP. The first mode is called "Regulation Mode of Compensating Jingjiang river Section", with an aim to control the water level in Shashi. It is suitable for floods coming mainly from the upper reaches of the Yangtze river, and is efficient in preventing special great floods as the ones in 1860 and 1870. This kind of simple and very operational regulation can successfully meet the flood control requirements from the middle and lower reaches of the Yangtze River. It is also useful in reducing diversion flow to the areas near Chenglingji, but not so ideal. The second mode is called "Regulation mode of Compensating Chenglingji area", which gives considerations to both Jingjiang River section and Chenglingji river section. It is applicable to floods coming mainly from the middle and lower reaches of the Yangtze river as well as from the whole basin, like the floods in 1931, 1954 and 1998. With complex but very operational regulation, it can reach the aims of flood control, and especially it is better in reduction of diversion flow to Chenglingji area than the first regulation mode.

7. The Role of TGP in the Flood Control of Yangtze River

After its completion TGP will be the backbone project in the flood control system of the Yangtze River and benefit greatly to the middle and lower reaches. Especially it will bring great change to the flood control situation of the Jingjiang area.

7.1 Decisive Measure in Solving Flood Problem in Jingjiang Area

From past to now the section of the Jingjiang River has been the focus of flood control in the Yangtze River valley. Even though supported by embankments, its standard of flood control can only meet the 10-year frequency flood, and although the Jingjiang flood storage and detention areas are applicable, it can defend floods at most 80000m^3/s of Zhicheng with a return period of every 40 years, and there is no reliable countermeasure for floods beyond the standard. However, with the completion of the TGP, the water level in Shashi will not be more than 44.5~45m when the Jingjiang Area is hit by the 100-year frequency flood without using the Jingjiang flood storage and detention areas. With regulation of Three Gorges

Reservoir, the peak discharge in Zhecheng can be controlled to less than 80000m^3/s if it is hit by the 1000-year frequency flood or like that of 1870, with concurrent use of flood storage and detention areas in the Jingjiang Area, and both banks of the Jingjiang River won't be damaged. Disastrous damage such as a large number of deaths resulting from embankment-breaks won't happen.

7.2 Relieving the Worries about Flood Control in Wuhan City

With embankment protection belts, Wuhan, a big city at low altitudes, can endure floods occurring every 20-30 years. When floods above this standard occur, the integral flood control plan is needed, which diverts flood in the upstream and downstream of Wuhan. Additionally, there are still worries about the flood problems in Wuhan, if dike breaches occur in Jingjiang river, the flood will rush to Wuhan in a short way, and water levels in Wuhan will rise sharply. This is because the straight distance between Shashi and Wuhan is much short than that between the two regions along the river, and there are no effective barriers for flood control. Consequently, once the collapse, Jingjiang River dikes, enormous water will rush to Wuhan. The Three Gorges Project will remove the threats of dike breaches of the Jingjiang river, and then relieve the worries about floods in Wuhan.

7.3 Greatly Reducing Diversion Flow

At present, the main comprehensive measures for flood control of the middle and lower reaches of the Yangtze river include embankments, flood diversion areas, river course improvements, etc. Based on controlling floods like the one in 1954, flood diversion and detention areas with a capacity of 50 billion m^3 are arranged. Some of them are 5.4 billion m^3 in the Jingjiang area, 32 billion m^3 in the Chenglingji area, 6.8 billion m^3 in areas near Wuhan and 5 billion m^3 in areas near Hukou. After the completion of TGP, according to its regulation modes of "Compensating Jingjiang" or "Compensating Chenglingji" proposed in the preliminary design of TGP, there is still extra flow volume of 39.8 billion m^3 or 336m^3. This will greatly reduce flood detention capacity and then dramatically cut down the losses of the flood diversion.

7.4 Delaying Sedimentation in Dongting Lake and Keeping its Role in Regulating Floods

Dongting Lake is the watercourse and regulating reservoir, which receives water from Xiangjiang, Zishui, Yuanshui, Lishui rivers and a part of the Yangtze. Before the completion of TGP, Dongting Lake plays an important role in regulating floods, since by the mouths of Songzi, Fudu and Ouchi 1/3~1/4 of the peak flow discharge of the Jingjiang River flows into the Lake. However, a great deal of silt has been transported into the lake from the Yangtze river, which causes serious sedimentation and speeds up the area reduction of the lake. After the completion of TGP silt flowing down will be sharply reduced. Consequently silt flowing into the lake by the three mouths mentioned above will reduce drastically, and then its role of regulating floods will remain for a longer term.

Moreover, with the completion of TGP, rational flood control system for the middle and lower reaches of the Yangtze river will be formed by reservoirs, embankments, flood diversion and detention areas and river course improvements etc. As a result, reliabilities and activities of flood control regulation can be greatly strengthened, which will dramatically change the situations of flood control of the Yangtze river.

8. Conclusion

After the TGP is completed the middle and lower reaches of the Yangtze river will form a comprehensive flood control system. This system takes the embankment as its foundation, the Three Gorges Reservoir as its backbone and is complemented by flood diversion and detention projects, tributary reservoirs, river course improvement projects, conservation of water and soil as well as non-structure measures. Consequently, the capacity of flood control of the middle and lower reaches of the Yangtze river will be greatly enhanced, and the situations of flood control in the section of Jingjiang river will be dramatically improved. In this way catastrophic flood disasters can be avoided; the impact on society and economy by floods can be restricted; and deterioration of human survival environments can be mitigated.

On the other hand, the contradiction between the safe discharge of the river course in the middle and lower reaches or the Yangtze river and high peak discharge, and big flow volume of the river is very obvious, and the flood control storage of the TGP is relatively not enough. At the same time there is still 0.8 million km^2 catchment areas in the middle and lower Yangtze basin, which include main rainstorm areas like the Dabieshan mountain area, western Hunan to western Hubei area and area from Jiuling, Jiangxi to Yellow mountain, Anhui, etc. In the middle and lower Yangtze basin, rivers of Dongting Lake, Qingjiang river, Hanjiang river and other important tributaries run into the main Yangtze river. The flow volume is high and its components are complicated. The flood control situations in parts of the Yangtze river Basin is still very serious and the main stream of the middle and lower reaches will maintain high water levels if big floods like 1954 occur. According to the regulation mode of compensating Jingjiang or compensating Chenglingji proposed in the preliminary design of TGP, there will be extra volume of flood with 39.8 billion m^3 or 33.6 billion m^3. Hence, flood control of the Yangtze river still need comprehensive measures and constructions of flood control projects needs to be further strengthened. For a long period time, main constructions will still be embankments, flood diversion and detention areas, river course improvements and reservoirs. At the same time, non-structure measures for flood control need to be publicized.

In addition, the completion of the TGP will change the sediment and regime of the middle and lower reaches of the Yangtze river. Also new changes will arise about relations between detention and discharge of the main Yangtze river, about relations between rivers and lakes, and about the river morphology of the middle and lower reaches. Further studies are thus necessary to provide scientific supports for integrated development and regulation of the Yangtze river. The studies may cover:

- The regulation of the Three Gorges Reservoir.
- The relations between the Three Gorges Reservoir and the comprehensive system of flood control of the Yangtze river.

- The relations of the water levels and discharges of the main stream of the middle and lower reaches of the river.
- The relations between the rivers and the lakes.
- The relations between sediment movements and river bed evolutions, etc.

(Qihui Yu, Yongjun Zhang, Weiyu Tang are engineers at Changjiang Institute of Survey, Planning, Design and Research, Changjiang Water Resource Commission, China.)

12

Ethiopia's Water Dilemma

Uwe Hoering

Ethiopia is a land of hydrological contrasts. Its uneven, often unpredictable distribution of water greatly impacts its efforts to address poverty. Ethiopia has become something of a poster child for the dam industry, which contends that big dams are critical for ending its poverty. But most development analysts believe that the rural poor need smaller-scale water projects more suited to meeting their immediate needs. Uwe Hoering looks at the range of ways Ethiopia is harnessing its waters for poverty reduction.

Water in abundance. Over the Simian Hills in the highlands of Tigray it is raining heavily. In no time, there are large pools of water standing in the fields behind the low stone walls marking the plots. Dried-up erosion gullies turn into brownish torrents, and the Tekezze River down in the valley swells from a small, harmless and lazy stream into a wild monster of rapids, whirlpools and cataracts.

Water shortage. Every year in June, Tadesse Desta hopes for the timely start of the rainy season to start sowing *tef.* Only a few weeks later he has to worry that the grain will dry up and there will be another failed harvest. For at least six months of the year he and his family are dependent on food aid, like one in four people in Tigray.

Earlier this year, agricultural advisers from the provincial government turned up and asked Tadesse Desta and his neighbors to "harvest the rains." For digging the pit in which the downpours would be stored, he got a few days' rations of maize. To coat the pit with a blue plastic sheet, he had to borrow money. The advisors told him that the small ponds, about eight by eight meters with a holding capacity of hardly more than 60 cubic meters, would help him save his crops if the rains should not be sufficient. Since the government made "rainwater harvesting" into an official program, setting targets for local officials to meet, the slopes in the highlands are now punctuated with thousands of these "household tanks."

Unfortunately, these ponds are of little use except as breeding grounds for mosquitoes. They are too small to really save the harvest in case of emergency. Tadesse Desta doesn't know how he will ever pay back the loan for the blue plastic sheet.

The flaw in this program does not mean that all rainwater harvesting programs will fail at helping Ethiopia's rural poor improve their lives. Nor does it lend support to the arguments now being put forth by the Ethiopian government and the World Bank that large-scale dams will solve the nation's woes. But it does reveal that even well-intended programs aimed at reducing rural poverty can miss the mark substantially if they are poorly conceived.

The Under-Dammed Argument

"Ethiopia is poor because it doesn't use its enormous water potential," claims the World Bank. True, in the mountainous regions, precipitation is normally quite high. It is also true that most of the waters there rush down the steep slopes unused, dragging along vegetation, soil, stones, roads and bridges, digging ever more and ever deeper gullies, and causing heavy floods in the lowlands.

Coming with the tags "economic growth" and "poverty reduction," the World Bank's solution for Ethiopia is: Dams would control the floods and utilize the abundant waters for energy and agriculture. Therefore, whoever builds them should be considered "holy men," says John Briscoe, until recently the Bank's senior water adviser. For new hydraulic infrastructure, the Bank and other multilateral finance institutions are willing to lend billions of dollars to Ethiopia, one of the poorest countries in the world.

The Bank's new Country Water Resources Assistance Strategy for Ethiopia says that to develop water storage capacity "must be seen as a development priority across the entire country" in order to improve water availability year-round. With just 43 cubic metres of storage capacity per capita, Ethiopia is far behind South Africa, whose 750 cubic metres of storage capacity per capita is being put forth as a rough water-security standard by the World Bank. The cost of attaining the "South African standard" is estimated at US$35 billion – five times the current gross national product of Ethiopia. Because of the "far reaching potential benefits of multi-purpose dam development, and the unique qualifications of the Bank to support these investments," the Country Strategy argues, a first priority for future Bank assistance in water resource management should be support for large dam development and river basin water transfers in the Nile River Basin.

The idea itself is not so new. As early as the nineteenth century, Ethiopia's Emperor Menelik II had plans to divert the Blue Nile from its deep gorge into the arid, sparsely populated lowlands in the western part of the country. Since then, time and again governments in Addis Ababa have devised elaborate plans for dams and irrigation projects. But again and again, Sudan and especially Egypt have managed to torpedo the implementation of these projects with diplomacy and military threats. They were afraid that any diversion of water in the upper regions of the Nile would negatively impact their own farmers downstream and their expansionist plans for huge new plantations and settlements in desert areas.

Nile Basin Initiative is Born

The pressure worked on the World Bank too. As part of an effort to get consent from Khartoum and Cairo for funding new dams on the tributaries of the Nile in Ethiopia, Uganda and Tanzania, the Bank spearheaded the Nile Basin Initiative about 10 years ago, functioning as the donor coordinator while the Ethiopian government serves as the facilitator. Progress to bring the riparian states together for an integrated river basin management plan benefiting all member states has been slow. But recently, the Council of Ministers accepted four hydropower and four irrigation development projects proposed by Ethiopia.

The biggest and most ambitious ongoing project is not part of this initiative. It is fundamentally altering a remote area at the tail end of a Africa's deepest canyon (2,000 meters deep in places) cut by the Tekeze River. In this stunningly beautiful canyon, construction is underway on a huge dam. At 185 meters, Tekeze Dam looms 10 meters higher than the gigantic Three Gorges Dam on China's Yangtze River. Tunnels several kilometers long are being driven through the rocks, and will divert the flow of the Tekeze into a huge reservoir, generating 225 MW of power, thus increasing Ethiopia's installed capacity by nearly one-third.

Because Addis Ababa became impatient with the slow pace of negotiations at the Nile Basin Initiative, four years ago the government decided to go it alone on this project. There were no consultations with neighboring Sudan, nor with its long-time foe Eritrea, which would like to use the border river itself. In far-away Beijing, which is systematically building up its engagement in Africa, Addis Ababa found a sympathetic financier for the $224 million project. The state-owned China Water Resources and Hydropower Engineering Corporation (CWHRC) not only undercut all other competitors, but also offered valuable experience with mega-projects because of its involvement in the Three Gorges Dam. "Tekeze Dam is for Ethiopia what Three Gorges is for China," claimed Sun Yue, Director of the international department of the CWHRC, at the contract signing ceremony. And it looks as if this is only the beginning. In July it was announced that China's Gezhouba Water and Power Co., is likely to win the contract to build a 100 MW hydropower dam on Ethiopia's Neshi River.

Although the World Bank has concerns about this bilateral move, economically, it thinks along the same lines: Tekeze's power could help Ethiopia's economic growth; the water storage would reduce the threats of floods and create opportunities for an intensification of food production. David Grey, the World Bank's Senior Water Advisor for Africa, contends that large-scale dams like Tekeze would be to the advantage of Ethiopia's poor. "There is no precedent for a country developing without harnessing its rivers and utilizing its water resources," says Grey.

But there is a fundamental flaw to this argument. This dam's power will go mainly to the cities or will be sold to neighbors with more developed industrial economies, and the water will irrigate fields downstream in the lowlands. But the poor – like Tadesse Desta, who year after year, are in need of food aid – live in the densely populated highlands far above the dams. The expansion of irrigation will only benefit richer farmers and foreign-owned plantations, because they have the influence and the money to make use of the new opportunities, developed with public money. Such developments also don't mean that there will be more food, because the production of low-priced food crops for local markets is not considered economically viable against the cost of new large dams. Instead, water and newly reclaimed lands will be used for the production of flowers, fruits or spices for export, or for cotton and sugar cane – water for cash and profit crops, not for food crops.

Neither Too Big nor Too Small

The dam near the village of Adis Nifas, not far away from the provincial capital Mekelle, is meeting quite different needs from Tekeze. The medium-sized structure (around 15 meters high and 300 meters long) has been built by the people themselves with the support of the Relief Society for Tigray (REST), a parastatal development agency. REST provided machinery and money, the villagers their labor. The reservoir is close to the fields, and most of the material used is local stones and sand, apart from some cement for the overflow canal and some valves and pipes to regulate the flow. Six months after the end of the rainy season, there is still a shallow pool of water left in the reservoir.

In the valley below the dam, each family received a quarter of a hectare of irrigated land, fruit tree seedlings, and elephant grass to plant on the earthen walls dividing the fields to reduce soil and water erosion. A water-users association has been formed to manage the distribution of water from the dam and maintenance of the small canals. Improved cultivation methods were introduced, which bring higher yields while also saving scarce water resources. Some families try their hands at cotton, sesame and vegetables. But most of them planted chilies, which are in high demand. Hardly any Ethiopian food goes without *Berbere,* the hellishly hot red paste.

Farther below the irrigated fields, the groundwater level is rising. From open wells 4-5 meters deep, farmers can lift water with a treadle pump onto their small fields. The longer the additional water is available beyond the rainy season, the better the crops and the higher the yields. Increased incomes would allow for further improvement in cropping methods, for improved seeds and fertilizer, but also for school fees, a new iron sheet for the hut, or a radio. "Tigray is not a hopeless case," says REST's Mulugeta Berhanu.

The priority for water development in Ethiopia should be many thousands, even tens of thousands, of small and medium sized dams like the one in Adi Nifas, says Helmut Spohn, who has been assigned by the German funding agency, Bread for the World, to assist small farmers in Ethiopia. The dams should be accompanied by afforestation, gully plugging and terrassing of the hills to avoid further erosion of the remaining soils. That would allow the rains to seep into the ground and recharge groundwater and aquifers, which still are the best and cheapest water storage, releasing it slowly over time, giving new life to perennial streams. It would also stop soil, sand and stones from being washed into the rivers with every rain.

Without such a program, the new mega dams and their reservoirs will be silted up after a few years. The result would be less additional power than calculated, less irrigation, less economic growth and less foreign exchange for the government in Addis Ababa. By that time, the consultants and construction companies would have been long gone, with their booty. But the government would still be sitting on its debt with the Chinese government or with the World Bank – rather like Tadesse Desta, who still owes the local money lender for the blue plastic sheet on his own ill-conceived water harvesting project.

(Uwe Hoering works as a freelance journalist on development issues, with a current focus on water and agriculture.)

Photo: RAIN

Ethiopia's water variability and lack of public development of water resources leaves farmers with shriveling crops in drought times, and women struggling to get enough water for their families' needs on a daily basis. The group RAIN with local partner Water Action is building rainwater harvesting tanks in dryer parts of the country. The project has benefited 3,000 people so far. Here, people line up for water at a Water Action-built tank.

13

Flooding the Future: Hydropower and Cultural Survival in the Salween River Basin

Burma is considered to be among most poorly governed states thus human rights violation and environmental impacts due to construction of large dams on Salween river basin have become a matter of concern. Also Burma will not be benfited from these hydropower projects as the electricity will be transmitted to Thailand and Vietnam. EarthRights International has given recommendations regarding hydropower projects to State Peace and Development Council (SPDC), multilateral banks, ILO and NGOs.

In April 2004, Thailand's Energy Ministry and Burma's Ministry of Electric Power agreed to develop four of the proposed projects. Joint feasibility studies began this past fall, prompting representatives from several different Burmese ethnic groups to urge Thailand to reconsider.

"If the dam is constructed blocking the river, not only will the Salween River stop flowing, but so will Shan history. Our culture will disappear as our houses, temples, and farms are flooded."[1]

– Shan refugee (2000).

"If the Wei Gyi Dam is built, it will not only stop the river. The Karen and Karenni will lose their homeland, farmland... and culture. The dam will only support the regime, not the indigenous peoples of Burma."

– Shwe Maung, a Karen man working in Thailand (2004).

These fears are on the verge of becoming true. After years of speculation, the Royal Thai Government and the State Peace and Development Council (SPDC), the military regime ruling Burma, appear poised to begin major construction on a series of large hydro-powered dams in the Salween River basin.[2] In April 2004, Thailand's Energy Ministry and Burma's Ministry of Electric Power agreed to develop four of the proposed projects. Joint feasibility studies began this past fall, prompting representatives from several different Burmese ethnic groups to urge Thailand to reconsider.[3] Their concerns emphasized the environmental costs of these dams and the fact that electricity produced from them would be exported abroad instead of catering to local populations who endure serious energy shortages.[4] Their pleas appear to have fallen on deaf ears.

Fortunately, the arrest of Khin Nyunt in October 2004, has fomented rather than ended the power struggle between different factions within the SPDC. As a result, the ongoing political turmoil inside the country has created a brief window of opportunity. But constructive action is needed quickly.

In December of 2004, newly appointed Prime Minister Lieutenant-General Soe Win received members of the International Monetary Fund (IMF), the World Bank (WB), and the Asia Development Bank (ADB) in Rangoon.[5] While the details of the meeting are not currently known, it is widely rumored that these multi-lateral development banks are considering renewing aid, especially humanitarian forms of assistance, to the country.[6] Should such aid begin, there will be little international leverage to stop or to modify plans for building large-scale hydropower dams along the Salween river.[7]

EarthRights International has conducted fact-finding on such development projects in Burma since 1996. Given our past findings – the Yadana/Yetagun Natural Gas Pipeline being just one prominent example – the dams are likely to result in severe human rights abuses and environmental degradation. This article will summarize the findings of several recent studies on this highly contentious issue, and offers a series of recommendations for addressing them.

Background

Burma's potential capacity for generating hydroelectric power is vast, though still largely untapped. Despite the military regime's human rights record—Burma is amongst the worst governed and most repressive countries in the world today—international financial institutions and energy companies have become increasingly aggressive in the hope, of exploiting this resource. Dozens of large-scale dams (more than fifteen meters in height) have been already built or are currently under construction throughout Burma, especially in the central region of the country.[8] Existing hydroelectric dams now produce about one-third of the country's electricity, though demand continues to far outpace supply in most parts of the country.

Feasibility studies are presently being conducted for many more dams, most of which are slated to be built in border regions, such as eastern Burma, where gross human rights abuses remain commonplace.[9] Alarmingly, many of these projects appear designed to supply non-local consumers as part of the Asian Development Bank's proposed "Mekong Power Grid."[10] The Mekong Power Grid is intended to supply power to urban areas, especially in Thailand and in Vietnam. However, there is little in these plans to suggest that the Mekong Power Grid will actually promote more sustainable forms of economic development throughout the region.

For example, the Tasang Dam in southern Shan State, if completed, will have a projected installed capacity of 3,500 megawatts, an amount which is three times Burma's current levels of electricity consumption. But nearly all of the electricity generated from the Tasang Hydropower Project is to be sent to Thailand via high voltage transmission lines, even though it currently enjoys an energy surplus. Indeed, the country's power reserve was as high as forty percent according to the Electricity Generating Authority of Thailand (March 2003 figures).[11]

Again, such figures undercut arguments that the Tasang Dam, much less a half-dozen of them along the Salween River Basin, is urgently needed to meet projected shortages. Rather, the proposed dams reflect the agendas of political, economic, and military elites and their perceived needs. While the ethnically diverse communities in Burma, whose livelihoods and cultural survival depends upon free-flowing rivers will suffer the costs of hydropower development, and receive few if any of its benefits.[12]

Damming the Salween

The Salween river is the longest free-running river in mainland Southeast Asia. The river originates on the Tibetan Plateau in the Himalayas and runs south through Yunnan Province of China. After entering Burma, the Salween river flows through the Shan, Karenni, Karen, and Mon States in the eastern part of the country before reaching the Andaman Sea. The river's total length is approximately 2,800 kilometers, second only to the Mekong river. Currently, eight dams and water diversion projects are planned on sections of the Salween river in Burma alone. Another series of large-scale dams, thirteen in total, are slated to be built on sections of the river further upstream in China, even though the area has been designated a "World Heritage Site" by UNESCO due to its rich biodiversity.[13] In the past, large infrastructure projects in Burma have led to grave human rights abuses committed by the increasing presence of military units providing security for the projects. The Yadana gas pipeline in Karen State is a prime example of how militarization led to earth rights abuses, which eventually were taken to trial in the US courts under the Alien Tort Claims Act.

> "*I would like to encourage foreign countries to think about how much suffering and destruction the building of the dam would cause for us. Without the dam, we already face many problems and struggle to survive. If the dam were to be built, I don't think we will be able to survive anymore.*"
>
> – Elderly Karen Villager.
> *(September 2003)*[14]

At present, ERI is most concerned about the following dams: i) Tasang; ii) Weigyi; and iii) Dagwin, as some preparation for construction has already

Potential Dam Sites on the Salween River

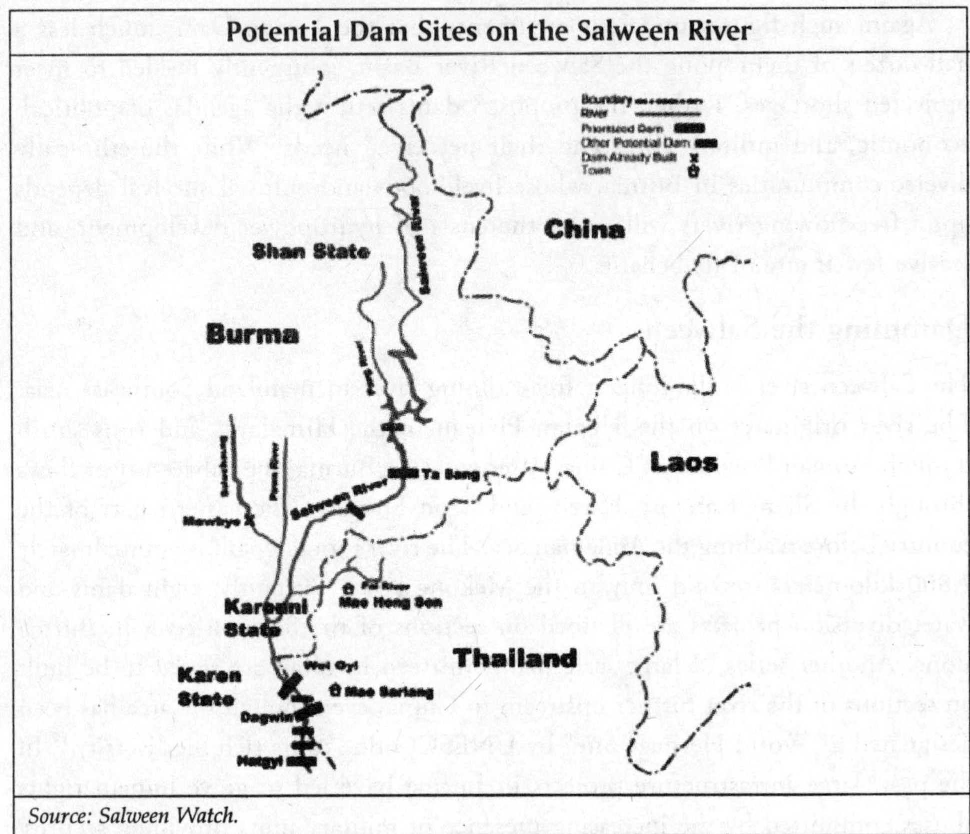

Source: Salween Watch.

Details of Planned Dams[15]

Name of Dam	Location in Burma	Megawatts (Estimated)	Status
Name unknown	Shan State	3,200	Proposed
Upper Tasang & Lower Tasang	Shan State	3,300-3,600	Feasibility Study Complete; Some Construction Underway
Name unknown	Shan State	4,000	Proposed
Weigyi	Karen State	4,540	Preliminary Studies Complete; Preparing Construction
Dagwin	Karen State	792	Preliminary Studies Complete; Preparing Construction
Hutgyi (a)	Karen State	5,850	Proposed
Hutgyi (b)	Karen State	6,000	Proposed
Hutgyi (c)	Karen State	10,000	Proposed

begun. To date, public participation has been completely absent surrounding these three dam projects, and any impact assessments, if conducted, remains closely guarded secrets. Importantly, the three dams are intimately connected to ongoing counter-insurgency campaigns carried out by the Burmese military in Shan, Karenni, and Karen States. ERI has gathered eyewitness as testimonies that indicate human rights abuses are widespread in the areas surrounding these three sites as well. Forced labor and portering have already been linked to the three dams and their security.[16] A summary of the situation follows:

Shan State

- *Militarization:* Since 1996, the Burmese military has implemented a forced relocation program of unprecedented scale to end support for the Shan State Army. Numerous studies have documented the Burmese military's deliberate use of rape and sexual assault against Shan women as well as women from other ethnic groups, which has been a central component of counter-insurgency strategy.[17] Currently, at least seventeen army battalions are now based in the area around the Tasang Dam site according to local sources.[18]

- *Depopulation:* Over 365,000 people have been forcibly relocated either to or in Shan State, many of whom now live in 176 relocation sites controlled by the Burmese military. Shan State also has the largest number of internally displaced people in the country; recent estimates place the figure at 275,000.[19] At least 2,000 of these households lived near the Tasang Dam site, prompting charges that they were moved so the SPDC would not have to pay land compensation claims later.[20]

- *Catastrophic collapse:* The concrete-faced rockfill dam would be 188-193 meters high, making it the highest dam in mainland Southeast Asia. The reservoir is expected to flood an area of at least 640 square kilometers.[21] The sheer size of both, raises the possibility of collapse, as the dam is built in a region where earthquakes regularly occur.

- *Greater food insecurity:* The same reservoir will flood productive farmland in low-lying valleys. Such arable land is in very scarce supply in Shan State due to its mountainous terrain. A considerable portion of the water, which will hold one-third of the Salween River's average annual flow, is also likely to be diverted directly to Thailand.[22] The reduced flow will adversely affect Burmese communities living downstream in a variety of ways.

Karen State

- *Militarization:* Between 1992 and 2004, the number of army garrisons in Papun District, where the Weigyi Dagwin Dams will be located, increased from ten to fifty-four, of these, a dozen are located around the dam sites.[23]
- *Depopulation:* Between 1992 and 2004, 210 villages in Papun District have been destroyed. Of the 85 village in areas close to the two dam sites, only a quarter remain. Former residents were forcibly moved to one of 31 relocation sites controlled by the army, where forced labor and other human rights abuses are common. Alternatively, they became an Internally Displaced Person (IDP) trying to survive in the jungle, or fled to Thailand.[24]
- *Increased Environmental Degradation:* New road construction to the dam sites will facilitate illegal logging, gold mining, and hunting in the Upper Mae Tha Lot Forest Reservation, the Dar Gwin Forest Reservation, and the Dar Gwin Wildlife Sanctuary.[25]
- *Greater food insecurity:* The reservoir from the Weigyi Dam alone is expected to stretch 380 kilometers upstream, which will force villages in Karen State and Karenni State to relocate. More seriously, the reservoir will inundate one of the two main wetrice producing areas in Karen and Karenni States. Given the loss of this important food source and the large numbers of troops in the area, food security for ordinary people, especially Karenni, is expected to decline significantly.[26]

Over the past several years, the International Labor Organization (ILO) has taken unprecedented action against the SPDC in the effort to curtail the continued use of forced labor by the Burmese military, but to little effect. Given the regime's abysmal record on this issue, the increased militarization of the areas surrounding the Tasang, Weigyi, and Dagwin Dams, suggests that history is likely to repeat itself. But this time, the scale of the abuse is likely to be far larger than seen in the past.

If the projects go ahead as currently planned, tens of thousands of people will need to be relocated. People currently living in forced relocation sites under military control near the sites are also at great risk. Many of them are likely to be forced to help construct a dam that will provide them with little or no material benefits. Credible accounts of rape, torture, and extra-judicial killings, which are common now, will also increase.

Conclusions

While hydropower projects have brought economic benefits, they have also adversely affected millions of people worldwide who depend upon rivers for their survival. These projects have irreversibly damaged ecosystems and led to the loss of livelihoods, cultures, and the rights of populations displaced by dams. All of the dam projects proposed for the Salween River basin in Burma fail to meet the standards established by the World Commission in Dams, particularly those related to open and transparent decision-making.[27] In every instance, advocates for the dams have failed to include the affected communities in the decision-making process, which raises concerns that profits are again being put before local interests and needs.

The projects also fail to meet the basic principle of distributive justice, which is embedded in the notion of sustainable development and other rights-based approaches. Sustainability, according to the 1980 World Commission on Environment and Development, cannot be achieved if policies do not consider the ramifications of resource accessibility and the equitable distribution of benefits and burdens across all affected stakeholders, including non-human ones.[28]

Current conditions inside Burma do not permit any of the above principles to be honored.

For these reasons, further construction should be halted until other, less destructive options can be explored, discussed, and agreed upon by all the stakeholders.

Recommendations

To the State Peace and Development Council (SPDC)

- Release Daw Aung San Suu Kyi and other members of the National League for Democracy who are still under house arrest or in prison for political reasons.
- Begin a credible tripartite dialogue to create a legitimate roadmap for democratization with specific criteria, timetable and milestones for measuring progress.

- Review and revise criminal laws listed above relating to freedom of nonviolent expression and association as defined in Articles 19 and 20 of the Universal Declaration of Human Rights (1948).
- Accede to the following international human rights treaties: the International Covenant on Economic, Social and Cultural Rights; the Convention against Torture and other cruel, inhuman or degrading treatment or punishment, and its optional Protocol as well as the ILO Convention on Forced Labor No.105.
- Require social and environmental impact assessments to be conducted by a qualified and neutral agency for all large-scale development projects.
- Conduct all matters related to the development of hydropower in a transparent and open manner, which includes the meaningful involvement of local people in the decision-making process.

To the International Labor Organization (ILO)

- Call its members to review their relations with Rangoon, especially which engage the regime through their involvement in hydropower development. Pressure the following to cease providing technical and financial assistance to the dams: Nippon Koei (Japan), World Impact Co., the Electric Power Development Corporation (Japan), the Japan Bank for International Cooperation, the Japan International Cooperation Agency, the Electrical Generating Authority of Thailand, and the Asian Development Bank.
- Call on international bodies, such as other UN agencies, and multilateral development banks to review their activities in Burma.

To Multi-Lateral Banks

- The Asian Development Bank should cease providing technical assistance to Burma via the Greater Mekong Subregion Scheme and soliciting funds connected to the Tasang Dam. Additionally, the Asian Development Bank should immediately expel the regime's representative serving on its Board of Executive Directors.
- The World Bank should not engage with the SPDC nor support development projects in Burma due to the military regime's history of human rights abuses, including the pervasive use of forced labor.

- International financial institutions should reject all grants to develop large-scale hydropower development in Burma until democratic rule is restored and mechanisms are in place for ordinary people to participate in decision-making, which would include a functioning framework for transboundary watershed governance.
- Any humanitarian assistance awarded to the country should not include funds or technical assistance for large-scale hydropower development in Burma.

To Non-Governmental Organizations (NGOs)

- Publicly support the ILO towards taking the above steps.
- Apply pressure on government agencies and companies in Thailand and Japan, urging them from moving forward with the proposed dams on the Salween River.
  - Lobby elected officials in your countries of origin to lodge diplomatic protests via formal government channels and the United Nations.
- Work in collaboration with Burmese organizations and civil society groups to develop alternate plans to large-scale dams. Such plans should:
  - Promote the more effective utilization of existing power plants.
  - Seek to reduce overall power consumption.
  - Improve demand-size management; to develop more "clean" energy through small-scale projects (e.g., biomass, wind, solar, geothermal, rainwater harvesting techniques, and mini-hydro power).
  - Decentralize water management through the greater use of small-scale reservoirs and forms of water storage.
- Prevent wastage in existing irrigation systems by fixing leaks and limiting evaporation.
- Draft mechanisms for transboundary watershed governance that include economic and other instruments to promote sustainable use, legal and regulatory frameworks for allocating river rights, and so on.
- Prepare and disseminate information in local languages regarding legal rights.

(EarthRights International (ERI) is a 501(c)(3) nonprofit group of activists, organizers, and lawyers with expertise in human rights, the environment, and corporate and government accountability. ERI has offices in the US and Southeast Asia.)

Endnotes

1 EarthRights International (ERI), Fatally Flawed: The Tasang Dam on the Salween River, (Chiang Mai: ERI, 2001).

2 Pradit Ruangdit, "Ties with Burma: No Backing for Rebels, PM Tells Junta," *Bangkok Post* (26 August 2004).

3 Piyaporn Wongruang, "Plea to Stop Supporting Salween Dam Projects," *Bangkok Post* (11 November 2004).

4 The Baluchaung /Lawpita Hydro-power Plant in Karenni State offers an excellent example of how local populations bear the high costs of such projects but reap few of the benefits. All electricity from the site is sent to Rangoon and Mandalay. Salween Watch & et al., *The Salween Under Threat: Damming the Longest Free River in Southeast Asia* (Chiang Mai: October 2004), p. 40-1. RWESA, "Baluchaung Hydropower Plant No. 2 Proposed Principles and Scope for Human Rights and Environmental Survey to the Ministry of Foreign Affairs, Japan" available at *http://www.rwesa.org/document/baluchaung.pdf.*

5 Prime Minister receives officials of IMF, WB and ADB, (14 December 2004) available at *http:/ /www.myanmar.com/nlm/enlm/Dec15_h3.html.* For further background, see Bank Information Center, "Status of Burma at the MDBs," available at *http://www.bicusa.org/bicusa/issues/ status_of_burma_at_the_mdbs/index.php.*

6 Yuki Akimoto, "The Multilateral Banks and Burma," *The Irrawaddy* 2004 (12) 4 available at *http://www.irrawaddy.org/database/2004/vol12.4/guest.html.*

7 For example, in 2001, the Government of Japan provided a US$ 28.6 million grant to reconstruct turbines for the Baluchaung hydropower dam, arguing that the electricity it generated supplied hospitals in the country. The grant, which was the largest aid package since the 1988 crackdown on pro-democracy demonstrators, was widely seen to be a reward provided to the regime for renewing negotiations with Daw Aung San Suu Kyi. See Thomas Crampton, "Japan Rewards Burma for Political Opening," *International Herald Tribune* (26 April 2001). Credible reports indicate that forced labor remains common around the dam. See ERI, "Japanese ODA to Baluchaung Hydroelectric Power Plant in Burma," available here .

8 In recent years, over 160 dams have been hastily built across the country. This figure was reported in the *Myanmar Times* (2 October 2004). For a current map of hydropower projects in Burma, see *http://www.burmainfo.org/env/DamMapBurma.jpg.*

9 For a recent overview, see Amnesty International, *Myanmar: Lack of Security in Counter-Insurgency Areas* (London: AI, 2002).

10 For details on the Asian Development Bank's proposed Mekong Power Grid, see Bank Information Center, "Quiet but Steady: The Asian Development Bank's Support for Burma," available at *http://www.bicusa.org/bicusa/issues/misc_resources/1322.php*. See also International Rivers Network, "Trading Away the Future: The Mekong Power Grid," available at *www.irn.org/programs/mekong/030620.powergrid-bp.pdf*.

11 Cited in Salween Watch & et al., *The Salween Under Threat: Damming the Longest Free River in Southeast Asia* (Chiang Mai: October 2004), p. 73. See also ERI, "Fatally Flawed: The Tasang Dam on the Salween River" (2002) available at here; ERI, "Stop Construction of Planned Tasang Dam," (April 2004) available here.

12 For details, see Yuki Akimoto, "Hydro-Powering the Regime," *Irrawaddy* (30 June 2004).

13 On the proposed dams, see ERI's "Lancang/Mekong and Nu/Salween Rivers: Promoting Regional Watershed Governance Distributive Justice for Downstream Burmese Communities" (ERI Report, September 2004) available at as a pdf 197.35 Kb. On the UNESCO designation, see UNESCO. Decisions Adopted by the 27th Session of the World Heritage Committee in 2003. Available at *http://whc.unesco.org/archive/*decrec 03. htm# dec 8-c-4. For environmental details, see also UNEP and WCMC, "Three Parallel Rivers of Yunnan Protected Areas," (10 August 2004), available at *http://www.unep-wcmc.org/sites/wh/Three_Parallel.html*.

14 Karen Rivers Watch, *Damming at Gunpoint* (Kathoolei: KRW, November 2004), p. 68.

15 See Salween Watch, SEA Rivers Network, and the Center for Social Development Studies, *The Salween Watch Under Threat* (Chiang Mai, October 2004), p. 7.

16 See ERI, "Overview of the Planned Tasang Dam," available here. Evidence of similar problems in Karen State is submitted to the International Labor Organization on a periodic basis. These reports can be found at "Burma Project," here.

17 See, e.g., Human Rights Watch. *Out of Sight, Out of Mind: New Thai Policy Toward Burmese Refugees and Migrants* (New York: HRW, 2004); Shan Human Rights Foundation and Shan Women's Action Network, *License to Rape: The Burmese Military Regime's Use of Sexual Violence in the Ongoing War in Shan State, Burma* (Chiang Mai: SHRF and SWAN, 2002); B Apple and V Martin, *No Safe Place: Burma's Army and the Rape of Ethnic Women* (Washington D.C.: Refugee International, 2003).

18 An army battalion usually consists of five hundred troops in Burma.

19 Burmese Border Consortium (BBC), *Internally Displaced People and Relocation Sites in Eastern Burma* (Bangkok: BBC, September 2002), available at http://www.ibiblio.org/obl/docs/*BBC_Relocation_Site_Report_(11-9-02).htm.*

20 Salween Watch & et al., *The Salween Under Threat* (Chiang Mai: Salween Watch, October 2004), pp. 47-49.

21 C Vatcharasinthu and M S Babel, *Hydropower Potential and Water Diversion from the Salween Basin*, Paper presented from the Workshop on Transboundary Waters: The Salween Basin held in Chiang Mai, Thailand (September 1999).

22 Salween Watch & et al., *The Salween Under Threat*, pp. 65-69.

23 Karen Rivers Watch, *Damming at Gunpoint* (Kathoolei: KRW, November 2004), pp. 30-31

24 *Ibid.*, pp. 1, 34, 35.

25 Compare the maps in *Ibid.*, pp. 21, 23.

26 Compare the maps in *Ibid.*, pp. 55, 64.

27 World Commission on Dams, *Dams and Development: A New Framework for Decision-Making* (Earthscan Publications, 2000), p. 217. See also, "The World Commission on Dams: A New Framework for Decision-Making" available at *http://www.irn.org/programs/mekong/gmskit/01.wcdfactsheet.pdf.*

28 WCED, Our Common Future (Oxford: Oxford University Press, 1980; R Kuehn. A Taxonomy of Environmental Justice. Environmental Law Reporter (2000).

For further Information on the Salween River

Burma Issues
http://www.burmaissues.org/

Friends Without Borders
http://www.friends-withoutborders.org/

International Rivers Network
http://www.irn.org/

Mekong Watch
http://www.mekongwatch.org/

Salween News Network
http://www.salweenews.org/

Salween Watch
http://www.salweenwatch.org/

Southeast Asia Rivers Network
http://www.searin.org/

Shan Herald Agency for News
http://www.shanland.org/

Shan Women's Action Network (SWAN)
http://www.shanwomen.org/

Karen Rivers Watch
http://www.freewebs.com/ka_rw2003/

14

Kalpasar: India's Most Ambitious Project

Prabha Shastri Ranade

Large dams are necessary for meeting Gujarat's increasing water needs of agriculture and industries. Kalpasar is Gujarat's key project after the completion of Narmada project. This project involves building a giant 64-km dam across the Gulf of Khambat from Ghoga in Bhavnagar district to Hansot in Bharuch district. The Gulf of Khambhat was identified as a promising site for tidal power generation by UNDP expert. This will help to trap the water from 12 rivers – including Narmada, Mahi and Sabarmati and create a huge freshwater lake. It will store three times the water in the Sardar Sarovar Reservoir, but it will not displace people. Through a 660-km canal system, 1.05 million hectares of land in coastal Saurashtra will be irrigated. It would provide about 5,461 million cubic metres (mcm) of irrigation water to Saurashtra – 900 mcm for drinking and another 500 mcm for industrial purposes. It will produce power to the tune of 6,000 MW through underwater turbines in the sea. A multi-lane highway and a railway line to be built across the length of the dam will reduce the distance between South Gujarat and Mumbai and also Saurashtra by 225 km. This project is to ensure a long-term water security in Gujarat with immense benefits is under criticism by the environmentalists.

Gujarat State: Geographical Background and Water Resources Scenario

Gujarat enjoys the natural advantage of location on the western side of the Indian peninsula. It has the longest coastline among the coastal states of our country, extending from Pakistan border to Maharashtra over 1650 km. Two of the three Indian Gulfs are also located in Gujarat. Gujarat has vast stretches of semiarid or parched land where there is scarcity or shortage of not only irrigation water, but also potable water for drinking. The average annual rainfall, restricted to only 3 months of the monsoon season, is 760 mm. It is very erratic. There are lots of variations in rainfall received in different regions of Gujarat. The southern part receives more rainfall up to 2500 mm, while the northern portion including Kutch and the Saurashtra peninsula receives very low rainfall of 300 mm to 450 mm. Thus the rivers in the state are small with very low and highly variable flows. Every third year is a drought year in Saurashtra, North Gujarat and Kutch region. Gujarat experienced severe droughts in the years 1987 and 2000. It has become necessary to store monsoon waters for dry period. Large areas of Gujarat have saline groundwater, which contains harmful proportions of fluorides and nitrates. The availability of usable groundwater is limited and that too has been overexploited over the past 3 to 4 decades. Every year groundwater tables are going down by about 3 meters in North Gujarat and Kutch regions.

A study on sources of economic growth and acceleration in Gujarat (EPW) reveals that the growth performance of Gujarat is better than the nation during the reform period (post-1991) in only eight sub-sectors. These sub-sectors include electricity, gas and water supply.[1] Gujarat government tried to get the maximum share of the Narmada waters (Sardar Sarovar project) and wanted to raise the height of the dam. After the completion of this project, the government claims that Gujarat will create a water revolution with no parallel in the country as it has utilized the Narmada water for reviving various rivers in Gujarat.[2] Soon, the Narmada water is expected to fill up the seasonal rivers like Banas, Rupen and Saraswati. A 700-km long pipeline is laid to carry the Narmada water to towns and villages in Saurashtra & Kutch region to quench the thirst of millions in drought stricken parched hinterlands. The Narmada water from Kevadia has reached Kutch after a 900-km long journey of Gujarat. It is a remarkable feat in civil engineering.

Kalpasar Project: Historical Background

Kalpasar is Gujarat's key project after the completion of Narmada project. The project has emerged as a mega-multi-purpose project. This idea was evolved way back in 1986 to solve the water woes of perennially parched lands of the state. "The name derives from the mythical concept of Kalpvriksha – a wish-fulfilling tree. The proposed project is that of a wish-fulfilling lake, therefore it is named Kalpasar." Former vice-chancellor of MS University, Dr Anil Kane first conceived the blueprint for the project 18 years ago. Anil Kane and Rajkot-based engineer V O Bhalodia are the brains behind the project whose conception began during the acute droughts of 1985, 1986 and 1987. The positive reconnaissance report of the project had come out in 1998.[3] The Gulf of Khambhat was identified as a promising site for tidal power generation by UNDP expert, Eric Wilson in 1975. In 1988-89 a reconnaissance report was prepared for the dam across the Gulf of Khambhat. The report concluded that, assuming sound foundation conditions, the closure of Gulf was technically feasible.[4]

Pre-Feasibility Report

The project's technical feasibility was established in a survey done by a Dutch consulting firm appointed by the State government, which was completed in 1998. This was followed up by six detailed surveys on various aspects.[5] National Institute of Ocean Technology (NIOT) run by the Central government was in charge of the Bathymetry survey of this project, which is essential before any drilling work can start on the seabed. NIOT under the Ministry of Earth Sciences has developed reliable indigenous technology to solve the various engineering problems associated with harvesting of non-living and living resources in the Indian Exclusive Economic Zone (EEZ), which is about two-thirds of the land area of India.[6] The NIOT has opened a centre at Bhavnagar only for this project. The bathymetry study cost about Rs.25 million and took about 3 months. A 12-member team using two tugboats sailed across the gulf at least twice a week to conduct the necessary surveys. Geotechnical studies were carried out to examine the seabed and findout whether it will be able to take the weight of the dam's pillars.

The study concluded that tidal power generation benefits should be supplemented by creation of a sweet water basin by impounding surplus water of the Narmada, Dhadhar, Mahi, Sabarmati rivers, which would provide irrigation, water supply and reclamation benefits. It concluded that the development of Gulf of Khambhat is feasible from the technical and socio-economic point of view.[7]

Government Approval

The Mission Statement on Gujarat government website mentions that Gujarat government is determined to ensure long-term water security in Gujarat in the new millennium by giving concrete shape to Kalpasar project. The Gujarat government has set up a special department for this purpose, namely, Department of Narmada, Water resources, Water Supply and Kalpasar. The government has approved the Kalpasar Project to provide new lease of life to the people of Gujarat, for providing water for irrigation to convert the semiarid north Gujarat region into a lush green belt. Gujarat government had approved REMAG project for restoration of mangroves around the Gulf of Khambat. It was found that 4,000 ha out of the total 5,000 ha of the REMAG project area coincided with Kalpasar project. REMAG had to revise its entire project area and had to identify fresh sites.[8]

Chief Minister Narendra Modi formally launched the Kalpasar project on February 5,2004 at Bhavnagar during a short voyage in the Arabian Sea. Modi pointed out, that this was the first time in 5,000 years that a sea would be dammed, after Lord Hanuman's monkey corps built a mythical bridge linking India and Sri Lanka, (Bharatiya Janata Party, Gujarat).[9]

The Gujarat government has decided to go ahead with the Rs.55,000-crore multi-purpose Kalpasar project as a long-term strategy to tide over the perennial water crisis, generate electricity from tidal waves and create road transport facilities. However, in spite of the government decision, the project would commence only in 2011 and would be complete by the year 2020[10]. According to other sources, the project is expected to be completed in 20 years period.

Salient Features of Kalpasar Project

Kalpasar will be a "dam with a difference" as it does not submerge any land, it does not involve any human displacement and it has no negative impact on the environment. Most importantly, it does not depend on cooperation from any other state (Source bjpguj.org). Kalpasar project in Gujarat envisages the construction of a 64.16 km long dam across the Gulf of Khambat (also known as the Gulf of Cambay) joining Ghogha in Bhavnagar district (Saurashtra) and Hansot in Bharuch district on the east coast of the Gulf. It aims at turning a part of the Arabian Sea into a freshwater lake. It will trap water from 12 rivers that flow into the Gulf – including Narmada, Mahi, Sabarmati and Dhader and create a huge freshwater lake. The reservoir formed will be sub-divided into (a) tidal basin and (b) fresh water basin. This would provide benefits of tidal power generation, as well as use of large quantities of fresh water. The project, when completed, will have a vast fresh water reservoir with gross storage of 16,791 million cubic meters of water, 64.16 km long road and 35 meter wide dam. It will store more water than all existing major, medium and minor dams in the state. It will store three times the water in the Sardar Sarovar Reservoir. The government claims that this project will not lead to any displacement like other dams or irrigation projects. The location of Kalpasar project is shown in the map of Gujarat.

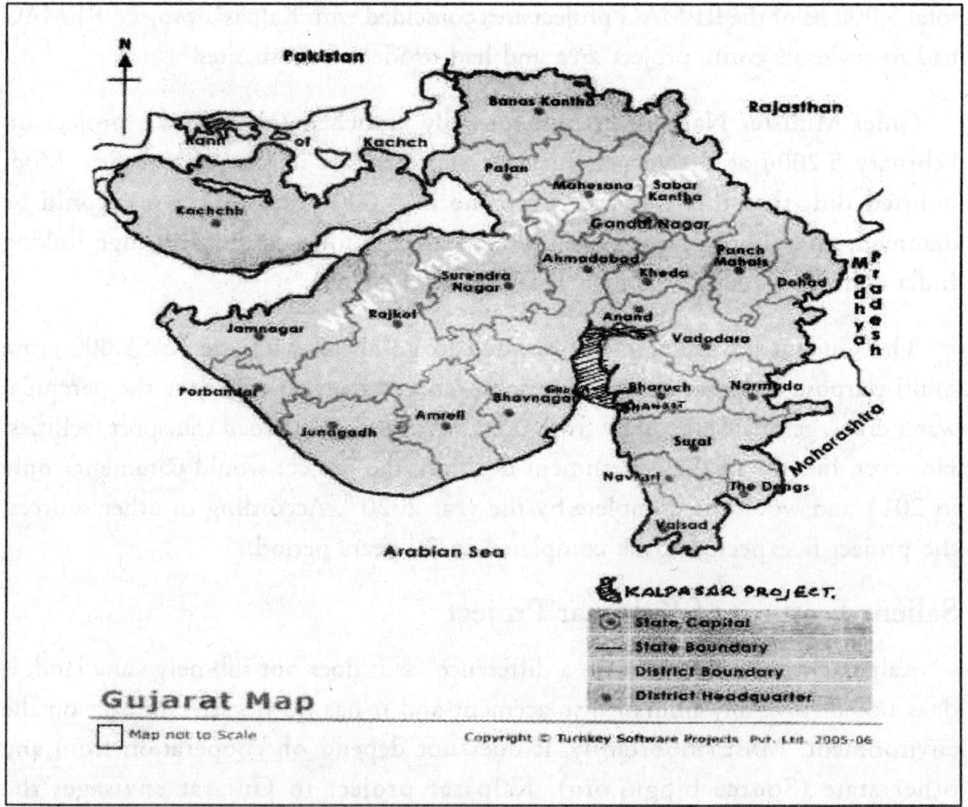

It will have a tidal wave power generation house with an installed capacity of 5,880 MW. The dam's top width of 35 meters will provide enough space for a road and a railway track. A multi-lane highway and a railway line is proposed to be built across the length of the dam. The project uses the simple concept of gradient differential between high and low tides which is among the highest in the world across the coast of Gujarat capable to produce power to the tune of 6000 MW, through underwater turbines in the sea.[11]

Similar Projects in Other Parts of the World

This unique freshwater reservoir is modeled on some of the existing ones in the Netherlands.[12] The Ijsselmeer project in the Netherlands has a 32-km long dam. The once marshy land in the vicinity has been used to create the city of Lelysted. There are similar projects like Kalpasar being implemented in other parts of the world. Some of which include France, Canada and Denmark. "Those projects, however, have focused only on water or energy, whereas we are aiming at an optimum balance between the two and also at various other benefits," Kane said. The La Rance project in France and the Annapolis project in Canada have attempted something along these lines. Other such projects are being implemented on experimental basis in the UK, Russia and China. However, the scope and size of the Kalpasaru project would perhaps exceed that of any other similar projects in the world.[13]

Scientific Studies for the Project

The government of Gujarat, after review of Pre-feasibility Report, decided to carry out six intermediate studies before a decision is taken on going ahead with the preparation of a full-scale Feasibility Report. These studies addressed some major concerns regarding environmental aspects and techno-economic feasibility as follows:

- Water quality in the future reservoir and sustainability of benefits.
- Hydraulic and morphological impacts within the reservoir, on the downstream as well as upstream.
- Drainage aggravation and salt balance on the peripheral area of the lake.

- Economy of tidal power generation and feasibility of absorption by India's Western Regional Grid.
- Review, reappraisal and integration of benefits from fresh water of Kalpasar reservoir with those of other projects in Gujarat.
- Economic and financial evaluation.[14]

The pre-feasibility study and six specific studies have shown that the project is technically and economically viable and detailed feasibility study needs to be taken up. A number of feasibility studies have been done by NEERI Nagpur, project management consultant. The institute of seismic research has given its report. A major programme of establishing comprehensive data bases for assessment of detailed environmental impact and other studies was launched and environmental clearance for the first phase of the project has been obtained. (*Times of India*, Ahmedabad, 17-03-07.) Swiss and French experts are involved in this project. They are carrying out detailed bathymetric and topographical/GIS surveys, geo-technical investigations, seismic studies, marine observations and analysis for the determination of design, height of the sea waves, silt content in tidal waters as well as the Arabian sea tides. A detailed feasibility report including a bankable project proposal will be prepared.[15]

The Benefits of Kalpasar Project

Kalpasar water will eventually lead to a quantum jump in living standards of the people in the region. Officials and those behind the project are of the opinion that the cost factor is not important keeping in view the returns of the project. This project is expected to solve the state's acute water problem. It promises a permanent solution for water scarcity in the perennially drought-prone Saurashtra region of Gujarat. It will provide 900 million cubic meters of water for domestic usage and 500 million cubic meters of water for the industrial development of Saurashtra and Kutch. It will provide around 5.61 million cubic meters of water annually to irrigate land of Southern Saurashtra, which is a water scarcity area. Through a 660-km canal system, 1.05 million hectares of land in coastal Saurashtra will be irrigated. The water will be available for domestic and industrial uses. It would provide about 5461 million cubic meters (mcm) of irrigation water to Saurashtra – 900 mcm for drinking and another 500 mcm for industrial purposes.[16] Around three lakh hectares of land in 13 talukas of Junagadh district would benefit from this irrigation project (*Times of India,* Ahmedabad, 17-03-07).

A multi-lane highway and a railway line to be built across the length of the dam will cut short the distance between South Gujarat and Mumbai and also Saurashtra by 225 km, saving an estimated Rs.8,680 million every year. The government has plans to build a road and a railway line on the dam. This road, connecting Talaja on Saurashtra side with Kim in south Gujarat will reduce the total distance between both sides from 380 km to 125 km.[17] With rising water level of the Arabian Sea, it will be possible to improve the existing ports like Ghoga and Bhavnagar. It is possible to build three to four new ports in the region because of the higher water levels. "The list of benefits does not end there. There is increased potential for port development. The state can develop the region as a tourist spot too."

The new fresh water lake can be used as a freshwater lake for fishery and this could generate an income of Rs.68 crore. This lake will help in reclaiming at least 1.2 lakh hectares of land. An additional 3 lakh hectares will be saved from salinity ingress.[18] Once the lake is built, the salinity level of the adjoining land will decline. About 1100 sq. km of land adjoining the Gujarat coast is saline and currently unfit for cultivation. Part of it can be reclaimed. The process of salinity ingress in the coastal regions can be stopped and land can be reclaimed. The project will generate 5880 MW of tidal power – more than the total capacity of the Gujarat Electricity Board.[19]

The following are the coastal talukas adjoining the Kalpasar project which will be benefited:

- Ghogha
- Bhavnagar
- Dhanduka
- Khambat
- Jambusar
- Vagara
- Bharuch
- Hansot

Criticism by Environmentalists

This ambitious project has been criticized by the environmentalists. They point out that the project may turn out to be a figment of imagination. They are taking objection for its construction on the ground that fishermen are likely to lose their livelihood and mangrove forests are likely to be destroyed. A combined report of NEERI, Nagpur and NIOT has identified 15 talukas of Gujarat as toxic hotspots. Some talukas are identified as major sources of sulphur and nitrogen oxides. Navsari, Pardi, Olpad and Mangrol have been identified as source of wastewater and solid waste. Many of the rivers in estuarine zone have a tendency to accumulate chemical waste as sufficient dilution and dispersion is not available. NEERI – NIOT survey has shown that the Narmada and the Tapi estuarine areas have been carrying so much effluent load and rendered the soil infertile, in turn.[20] The ecologists express doubts about the future of delicate marine ecology of Gulf of Cambay. They also point out the decreased flow in the downstream reaches of rivers flowing into Gulf of Cambay. What would be the impact of decreased flow in the downstream of Narmada on the Kalpasar is yet to be ascertained.

Financing and Future Prospects

Eighteen years back the estimated cost of the Kalpasar dam and tidal power works was estimated at Rs.44,301 crore and Rs.38,124 crore for single basin and double basin, respectively. The cost is approximately Rs.54,000 crores for single tidal basin options and Rs.47,800 crore for double tidal basin option at 1999 price level.[21] As the project involves huge expenditure, Gujarat government will have to explore innovative options to finance the project. Official sources are negotiating with international aid agencies including the World Bank, Asian Development Bank and Japanese Development Bank. Gujarat has successfully experimented in people's participation in developmental projects like building a series of check-dams. They hope to implement the Kalpasar project with people's participation. Anil Kane, the promoter of the project, has suggested that the project be privatized and a corporation called Kalpasar Development Authority be formed for the same. Major companies operating in Gujarat like Reliance, Essar, L&T, Tata have shown great interest in the project. (DNA Money, Ahmedabad, March 24, 2007).

Though the deadline for completion is 2018, it may not be possible to complete it by this deadline as the work is of gigantic proportion and there are many complications involved in building such a long dam in the deep sea. Minister incharge of Kalpasar department, Bhupendrasinh Chudasama in state legislative assembly had said that the state government has spent Rs.1,875 crore on the project by December 2002.[22] The same Minister while responding to a query in state legislative assembly on 16-03-07 said that till now Rs.2544.39 crore have been spent on various prefeasibility studies for the project (*Times of India,* Ahmedabad, 17-03-07). The state government officials believe that the project is not likely to be completed before 2020, while Dr Kane believes that on getting green signal from the state, the project will be completed within five to six years. The delay is also increasing the financial burden of the project. Experts say that the project cost would be recovered in 15 to 20 years, while the dam life is predicted to be 250 years.

(Prabha Shastri Ranade, Consulting Editor, Icfai Business School Research Centre, Ahmedabad).

Endnotes

1 Dholakia, Ravindra H, Sources of economic growth and acceleration in Gujarat, *Economic and Political Weekly,* March 3, 2007, p. 770.

2 *http://www.gujaratindia.com/Initiatives/Initiative1.htm*

3 ahmedabad.com 5 Feb 2004.

4 *http://guj-nwrws.gujarat.gov.in/kalpsar/default_e.htm*

5 *http://www.bjpguj.org/NEWS/2004/february/08022004_1.htm*

6 *http://www.niot.res.in*

7 *http://guj-nwrws.gujarat.gov.in/kalpsar/default_e.htm*

8 Source: REMAG-Restoration of Mangroves in Gujarat: A community based approach, Gujarat Ecology Commission, July, 2006.

9 *http://www.bjpguj.org/NEWS/2004/february/08022004_1.htm*

10 Source: Project Monitor, March 8, 2007.

11 Ahmedabad Newsline, March, 24, 2003.

12 Ahmedabad Newsline, March, 24, 2003.

13 *http://guj-nwrws.gujarat.gov.in/kalpsar/default_e.htm*

14 *http://guj-nwrws.gujarat.gov.in/kalpsar/default_e.htm*

15 *http://guj-nwrws.gujarat.gov.in/kalpsar/default_e.htm*

16 Ahmedabad Newsline, March, 24, 2003.

17 ahmedabad.com 5 February 2004.

18 Ahmedabad Newsline, March, 24, 2003.

19 *http://www.bjpguj.org/NEWS/2004/februaryl/08022004_1.htm*

20 *www.indiatogether.org/2004/mar/env-kalpsar.htm*

21 *http://guj-nwrws.gujarat.gov.in/kalpsar/default_e.htm*

22 ahmedabad.com 5 Feb 2004.

References

Kalpsar feasibility study gets back on track, February 5, 2004 Source: *ahmedabad.com*

Dholakia, Ravindra H, Sources of economic growth and acceleration in Gujarat, *Economic and Political Weekly,* March 3, 2007, p. 770.

Kalpasar project: feasibility study to be conducted, Express News Service, Ahmedabad Newsline, March, 24, 2003.

Kalpsar-NWRWS *(http://guj-nwrws.gujarat.gov.in/kalpsar/default_e.htm).*

Bhartiya Janta Party Gujarat *(http://www.bjpguj.org/NEWS/2004/februaryl/08022004_1.htm).*

Kalpsar feasibility study gets back on track (*http://www.ahmedabad.com/index/viewarticle/article/13025/section/14).*

Kalpsar project feasilbility study to be conducted *(http://cities.expressindia.com/fullstory.php?newsid=47169).*

Himanshu Upadhyaya, Kalpsar: a lake of wishes? *http:// www.indiatogether.org/2004/mar/env-kalpsar.htm*

Joydeep Roy, Modern India's most ambitious project, *http://in.redeff.com/money/2004/mar/06spec1.htm*

http://www.niot.res.in

Kalpsar project to commence in 2011 (*http://www.projectsmonitor.com/detailnews.asp?newsid=3407).*

REMAG-Restoration of Mangroves in Gujarat: A community based approach, Gujarat Ecology Commission, July, 2006.

15

Behind Bhakra, Beyond Bhakra*

Shripad Dharmadhikary

This article is the concluding part of a report "Unravelling Bhakra: Assessing the Temple of Resurgent India" which highlights the pros and cons of the Bhakra project which was the first of the storage dams in the Indus Basin. It briefly differentiates the diversion systems and the storage systems. It also draws attention to various exaggerated estimates proposed by Bhakra project and compares it with the findings of the author. It emphasizes the diminishing resources and the increasing disequilibrium in the natural cycles. Approaches for irrigation like soil water conservation measures and local rainwater harvesting have been suggested.

"More ... that is what we need. All the problems can be solved if we have more water."

– Comment by a participant in a public meeting in Narwana, Haryana

---

* The article is culled out from pgs.229-238 of the report "Unravelling Bhakra" authored by Shripad Dharmadhikary and publishe by Manthan Adhyayan Kendra.

"The proposed plan will not fully satisfy either side. No plan could do that; there is not enough water to fill all demands."

– Eugene Black, President of the World Bank, writing to Prime Ministers of India and Pakistan in 1954, urging them to accept the plan proposed by the Bank to resolved the Indus Water Dispute.

Behind Bhakra

Dedicating the Bhakra Project to the nation on 22nd October 1963, Pandit Jawaharlal Nehru was moved to say[1]:

"Bhakra Nangal is something tremendous, something stupendous, something which shakes you up when you see it. …"

It does. I still remember my first view of the dam – breathtaking, even overwhelming. Yet, for all this, our study found that the Bhakra dam and project to be a most ordinary project, an ordinary dam much like any other large dam – with all its flaws and blemishes.

We saw that the design of the dam was driven by the need to strengthen negotiating positions in the interstate disputes – first between Sind and Punjab, and later India and Pakistan – than the need to address the dry areas. This is a phenomenon that is seen in many other projects. We saw that while the justification being given was to take waters to the dry areas of Hissar tracts but the priority was given to augment the SVP. The areas proposed to be irrigated by the project had also been highly exaggerated – a familiar phenomenon in large dam projects.

Most dam projects, by the very fact that they store water that otherwise would have flowed on further down, work to transfer water from the downstream areas to the upstream. In the case of Bhakra, this was taken to the extreme, and the areas benefited by the project are actually a transfer of irrigation from downstream (SVP areas) to the upstream.

We found that the anticipated foodgrains production from the project – a crucial part of its very *raison dêtre* – had not been properly worked out but only some general estimates made.

In these and in many other ways the Bhakra project was just another dam. It was not much different as far as performance went. Indeed, its performance has been at gross variance with its larger-than-life public image.

We started with the widespread public perception that Punjab and Haryana are the granaries of the nation and that this is due to Bhakra. The "Punjab=Bhakra" (and to a lesser extent "Haryana=Bhakra") is an equation entrenched in popular mind in India. We soon found that this was far from the truth. Irrigation in the Punjab and Haryana had begun many decades before Bhakra. This was from diversion schemes including the Western Jamuna Canal, the Upper Bari Doab system, the Sirhind canals, to name only the major systems.

As far as Bhakra is concerned, 20% of the total cultivable area of Punjab is commanded by Bhakra. For Haryana, the same figure is 31%. *Punjab and Haryana are much more than Bhakra.*

One of the issues that this brings up is whether there is any difference between the irrigation from the diversion systems and storage systems. The advantage claimed for a storage structure is that it can provide better regulation, and especially help augment irrigation in winter, when river flows are smaller, by transferring excess monsoon water to winter months. But this is at the manifold costs including displacement, downstream deprivation and so on. A diversion structure, by its very nature, causes less disruption in the flow of the river. This is especially true of the monsoon or the high flow season. The weir or barrage will cause much less submergence and displacement as compared to a storage dam that creates a reservoir.[2] Thus, the major impacts in terms of displacement, submergence of forests, and severe downstream effects are avoided. There will be some impacts downstream – to the extent of the diversions taking place; but overall, the impacts are much smaller. The financial cost is also normally much less than a storage dam.

The irrigation developed in the Indus basin through these diversion schemes had much smaller social and environmental impacts. Much of the irrigation in the two states comes from such systems.

Even if we assume the contribution of an irrigation system to the production in the two states is in proportion to the area covered by it, Bhakra would not be responsible for more than 31% of Haryana's production and 20% of Punjab's. This is a far cry indeed from the public perception of Bhakra's role.

However, we found that this would be a gross misrepresentation. Between the two states, Punjab's production of foodgrains is twice as much as Haryana. And in Punjab, Bhakra has made little dent. An analysis of the command area reveals that much of the Bhakra command in Punjab was already irrigated, or was in wellendowed areas. And the irrigation from Bhakra canals played a limited role in these areas. In Punjab, irrigation development after the mid-1960s really took off with the explosive growth in the groundwater irrigation through tubewells. We may recollect here that the Bhakra project was designed to irrigate, at best, a maximum of 62% of the cultivable command area annually. In Punjab, even this was not achieved.

In Punjab, even in the Bhakra command areas, tubewell irrigation has been the overwhelming major source. *If there is one thing our study has exposed – it is that "Punjab=Bhakra" equation is a big myth.*

The growth of tubewell-based irrigation was mirrored in Haryana.

We saw that the real jump in foodgrain production came after the advent of the Green Revolution, with the coming of HYV seeds. While Bhakra is strongly associated with the Green Revolution in public mind, we need to note that neither the Green Revolution nor irrigation came to Punjab/Haryana with the Bhakra project. Irrigation was there over a hundred years before Bhakra, and the Green Revolution came in only 12 years after the irrigation from the project had begun. The Green Revolution is quite distinct from Bhakra.

There is little doubt, as our study clearly shows, that the driving force behind the Green Revolution was the tubewell-based irrigation. This was true of Haryana, and certainly of Punjab. The high rates of growth in the foodgrains production, and the cropping pattern that includes large area of rice could be achieved only by massive extraction of groundwater – far beyond the normal recharge.

So rapid has been the growth in the groundwater extraction that a huge part of Punjab and Haryana's production today comes from the areas dependent on unsustainable extraction of groundwater – 43% for Punjab and 34% for Haryana.

This is based on water, which is not being recharged, which had accumulated through decades or even centuries. *The miracle of Punjab and Haryana is, in reality, highly unsustainable, and now is in the process of a collapse.*

Our calculations show that the contribution of Bhakra project – including the benefit of the groundwater recharge due to the canals – is 11% in Punjab and 24% in Haryana. Compare this with the above-mentioned figures of production dependent on unsustainable mining of groundwater.

Our study has found that the impact of the Bhakra project was felt mainly in Haryana, that too in the drier districts of the Hissar tracts. The contribution of these areas – the areas served by Bhakra – has been limited. This limited contribution has come with huge costs. The costs of the dam – financial, social, ecological, the land degradation in the command areas, large scale waterlogging and salinisation of the soil which seems very difficult, if not impossible to manage, the deprivation of the areas downstream, the displacement of thousands of people, the impact of the prolonged and extensive use of chemicals and so on. These costs have been enormous, long-term and in all probability irreversible.

What is perhaps equally important is that these costs are translating into serious economic problems for the agriculture of the two states, threatening its very viability. Yields are stagnating, and more important, more and more inputs are being required to get the same output. *Margins of farmers are being squeezed; grains are too expensive for the people of the country to buy.* As the Johl Committee Report points out[3]:

> *"India has accumulated huge stocks of foodgrains that are not finding market .....Although as per the nutritional requirements of the Indian population, these stocks may not be considered in excess, yet due to the lack of purchasing power with the poor, supply exceeds demand....On the other side, the farmers, especially the farmers in the surplus producing areas, are experiencing an economic squeeze due to the decreasing margins between their costs of production and the prices they receive. Punjab in particular is in catch twenty-two...."*

These declining margins have created massive indebtedness among the farmers, leading many to be caught in a debt trap. In several cases this has even led the farmers to the extreme step of committing suicide.

Neither the farmer is happy, nor does the consumer gain. Was this the desired goal of the project?

Some may be quick to argue as to what has this got to do with the project. Was the project responsible for all these problems? We would pose a counter question – how is it that the project did not prevent this? That such a situation has arisen in spite of the project?

Also, if irrigation from the project is glorified by pointing to the spectacular increase in the agricultural production, then it needs to be recognised that this production was made possible, among other things, by the heavy use of chemical inputs along with the HYV seeds, a policy of Minimum Support Price and large- scale assured procurement. It is a package that has worked together. Indeed, it is an important question whether without the kind of productivity that these chemicals brought in, the dam itself would have been financially or economically viable[4]. Hence, the effects of the extensive and intensive use of chemicals and also the erosion of bio-diversity due to the very limited variety of seeds being used are part and parcel of the total cost of doing business with such irrigation projects.[5]

After all, no one is interested in the dam for its own sake.[6] The dam is a means to using the water resources for development – ensuring at the minimum adequate and affordable access to food for people and a reasonable livelihood for the farmers. If the long-term impacts result in these very goals being negated, then this necessitates some rethinking.

Large storage dams with extensive canal networks are among the most expensive of irrigation systems. To justify such an expensive interventions, the returns should be as much or more. With the current paradigm, such high returns (in terms of food or agricultural production) are possible only with the massive use of chemical and other inputs. This has led to the problems of soil degradation, threatening the long-term well-being of the system. It is quite possible that cheaper means of irrigation (or, more generally, of increasing crop productivity) will not require such high returns to make themselves viable, and can thus manage with lesser inputs, leading to lesser economic and ecological problems. Since they would also be decentralised, they would address in a better way the problem of equitable distribution.

Will such systems meet the problem of food production? We will address this issue in the next section in detail; but it may well bear repeating that the issue to be addressed is not just of food production – but also of food security and access to food. In terms of all three, we have no doubt that other means are more effective than large storage dams.

It is clear that the early dam projects – taken up immediately prior to or just after independence – hardly looked at this issue of the economic viability. The First Five Year Plan notes[7]:

"A number of projects – some multi-purpose and others only for irrigation – were sanctioned soon after the end of World War II. On some of these, works were started before the completion of detailed investigations and of economic studies necessary to obtain a correct appraisal of the technical and financial aspects of the projects...."

K.N. Raj is even stronger[8]:

"There had been no appraisals, of this scope[9], attempted in India either for the Bhakra Nangal or for any of the other proposed investment schemes. The project reports, on the basis of which the investment decisions are taken, give certain standard technical details and some estimates in very general term of the probable effects on production; these are supplemented by surveys and reports in regard to particular aspects of the projects, but the information given is usually fragmentary and analysis of the data does not add to anything like economic appraisal."

So what does this mean – that such projects are never viable? That we should never build a (large) dam? This is a huge debate that is of enormous contemporary relevance and significance, but it is not our intention to go into it here.[10] We will touch this debate in one respect though – and that is, the use of the Bhakra dam project as a model to justify large dam-building programs elsewhere in the country. Proponents of large dams point to the spectacular success of the agriculture in Punjab (and to an extent in Haryana) and attribute it to the Bhakra project. This is then used as an argument to advocate, justify or otherwise push for other large dam projects. It is an argument that is brought into play to counter (wish away?) the adverse impacts of large dam projects. The Bhakra project, used as a proxy for the agricultural "success" of Punjab, is used as an argument to end all arguments against large dams. So entrenched is the perception of Agricultural Success=Punjab=Bhakra that this argument often succeeds.

Our study has shown that this argument is widely off the mark. The agricultural success of Punjab and Haryana has been a short burst of prosperity that is not only stagnating but is plunging into economic, ecological and social crisis. And even this short burst has had little to do with Bhakra. Hence, the use of Bhakra as an argument to justify other large dams is a highly specious argument.

At this point we may also mention that apart from this basic flaw, the use of Bhakra to justify other large dams is problematic also because it is often not recognised that Bhakra was built under circumstances very different from what we face now in other parts of the country. Let us recollect some of the more important circumstances that were unique to Bhakra.

First of all, it should be noted that the Bhakra project did not create any new irrigated areas; it simply transferred the areas being irrigated by the SVP in Pakistan to India. Bhakra could provide irrigation to Hissar tracts only by drying up the whole of Sutluj below Ropar.

Secondly, Bhakra was built in the days of the newly-independent-nation euphoria. This euphoria, and the accompanying outpouring of patriotic sentiment was to push aside many problems with the project. We have already seen the attitude and approach of the oustees who were ready to put up with many serious shortcomings in the resettlement program.

The lack of corruption, certainly in the resettlement process (reflecting in all probability the lack of corruption overall) – at least in the early days – was another rather unique condition – never again to be seen in India. These two sets of circumstances made building the Bhakra much easier as it helped push aside major issues and problems.

Thus, using Bhakra as an argument to justify more large dams is a seriously flawed argument. Yet, Bhakra has been thus used countless number of times without understanding the facts behind it. *Raag Durbari*, the hilarious and hard-hitting satire has captured this very well[11]:

> "........Iss desh ke niwasi parampara ke kavi hain. Cheez ko samajhne ke pehle we usspar mugdh hokar kavita kehte hain. Bhakhra-Nangal baandh ko dekhkar we kah sakte hain, "Aha! apna chamatkar dikane ke liye, dekho Prabu ne phir se Bharat-Bhoomi ko hi chuna"".[12]

Beyond Bhakra

The Bhakra project represents, in a way, the climax of irrigation development in the Indus basin. The fascinating course of irrigation development in the Indus basin began with the advent of the Harappan age, starting with the *sailaba* agriculture, the earliest inundation canals then needed to be re-built every year, the evolution of these canals to long channels with anelaborate network of distributaries, to the construction of permanent headworks to enable perennial irrigation.

With the headworks the system changed from being an inundation system to a diversion system, with better control on the diversions from the river. Each of these phases was marked by increasing abstractions from the rivers. Yet, for long, these were small enough not to cause any significant change in the river flows.

With the advent of the British era, developments took place at a rapid pace, and as more and more diversion schemes came up, an interesting phenomenon that was not in the picture so far made its appearance. For the first time, the abstractions from the rivers started reaching such a point that the areas lower down started feeling the reduction in flows. This typically manifested itself in disputes between separate political entities when the two areas fell in distinct political divisions – the Sind-Punjab dispute is an example of this. Words like 'upstream' and 'downstream' started to take on a different meaning. Areas downstream began to feel concerned that "their" flows were being taken away.

Still, the concern so far was limited to the "lean season" flow, as the existing weirs and barrages could divert only limited quantities of the monsoon flows. Much of the floods would pass over the weirs or barrages and flow on downstream. Of course, as the number of points where the river was "tapped" increased, a larger portion of even the monsoon flows began to be diverted.

Around this time, the idea of storing the monsoon flows began to be floated. It was an enticing idea – to those who saw the waters running "away" past them. Little thought was given, of course, to the fact that this water that was flowing past them, was flowing on to someone else. As technology made it possible to translate this idea into reality, the era of large storage dams began. Bhakra was the first of the storage dams in the Indus basin, soon to be followed by other like the Pong on the Beas, the Mangla on Jhelum and Tarbela on the Indus. The storage dams brought with them a quantitative change in the abstraction of waters from the rivers.

At some point along this evolution, the abstraction turned into exploitation. *At some point, to use a modern term, the system became unsustainable.* At what point do the withdrawals from nature start becoming destructive and detrimental? This is one of the most heatedly debated, most contentious issues of today. This is the issue that lies at the core of our study.

The progression of increasing withdrawals from the rivers in the Indus basin was paralleled by similar developments in other areas.

The use of the bucket and rope and the *shaduf* to draw waters from the wells gave way to the Persian wheel. More water could now be drawn from the ground. The advent of diesel and electric motor pumps and tubewells led to a huge jump in the capacity to extract groundwater. For the first time in human history, human beings had at their disposal the means to bring out water faster than nature was recharging it.

Storage dams put for the first time the capacity in human hands to dry up rivers.

Settled agriculture was a step ahead in "taking" from the soil as compared to mere hunting-gathering. Double cropping, and multiple cropping increased this. With the advent of HYV seeds, came hugely increased capacity to take up from the soils – to the extent that the nutrients contained in the soils were not enough to feed the "hunger" of these seeds. Heavy inputs of chemical fertilisers were necessary to make possible the high productivity of these seeds.

The progressively increasing withdrawals in all these systems are at the heart of the dramatic growth of agriculture in Punjab and Haryana. *All these systems have passed the point of sustainability.*

Every element in the "success" of the agriculture in Punjab and Haryana is based on over-extraction. The states are pumping more groundwater than is being recharged. The seeds are drawing more from the soil than there is; and the dams are diverting away more water from the rivers than they should – all of it at huge social, ecological and economic costs. Clearly, if we want that a vast nation be fed from a limited area – then we will have to extract far more from this small part. If we want to develop agriculture that is going against the grain of the geo-climatic make up of an area, we will have to provide the inputs externally.

When and why does the level of extraction become unsustainable? Clearly, there will not be one answer. No one person can set a limit and say that this level is okay and not beyond this. But there are some broad parameters that can be our guide in this.

Nature is designed with cyclic processes. Various elements go through a cycle, getting transformed, transported in the process, but coming back to the original state. The water cycle is well-known, as is the carbon cycle. These cyclic processes are highly interlinked and are in a state of dynamic equilibrium.[13] In contrast, most human designed processes are linear in nature – on one side are the inputs, which transform into outputs, and there are by-products. The outputs and by-products ultimately become "wastes" often creating serious problems of disposal. In nature, there are no wastes – because outputs or "by-products" of one process are inputs for the next stage of the cycle.

Human interventions in the nature often tend to disrupt the natural cycles. *We would say that the extent to which human interventions lead to deviations from these cycles and disturb the equilibrium is a good measure of unsustainability.* To ensure sustainability, on the other hand, we need to be as close to the natural cycles as possible.

If we look at the irrigation and agriculture development in Haryana and Punjab through this perspective, we can understand what is happening. The river that was earlier flowing into the sea is now being diverted somewhere else. In parts this water is now accumulating in the soil causing water logging. The groundwater that had been recharged since centuries is being taken out, with no replacement. The nutrients that are taken from the soil do not get back to it. Instead, we are pouring in chemicals, themselves extracted by disrupting other cycles.

All these have consequences that even now we are not fully grasped of. But those we *can* see are serious enough.

Some say that these are merely problems of management. Better management, better technology and more money can set these problems right. This approach is often called the 'technological fix' approach.

We see more fundamental issues at the heart of the problem[14] – the essential unsustainability of these unlimited extractions. We believe that solutions will need a shift in the way of working. The need is to address the root cause of the problem – namely, the shift away from and the disruption of the natural cycles. *In fact, this approach is not to be limited only to Punjab and Haryana, but should be the guiding norm all over the country.*

In case of irrigation, this approach would mean starting with soilwater conservation measures and local rainwater harvesting, as this is what would cause minimum disruption. Groundwater use would have to be limited to the amount being recharged – though the amount being recharged can be increased through several measures. In case of agriculture, this approach would mean organic agriculture, with minimum of chemical inputs. It would also mean diversity of crops, it would also mean agriculture that is in consonance with the geo-climatic set up of the area.

Would these imply only moderate increase in yields of foodgrains? Even if it did, it would not be of concern if this meant moderate increase over large areas. Of course the increase may not necessarily be moderate. People working on such lines have achieved yields that are remarkable. During our visit to Haryana, we visited Sukho Majri, a place that is now famous nationally and internationally for its rainwater harvesting and soil water conservation efforts. Sukho Majri was in complete contrast to what we had seen in rest of Haryana. We saw large variety of crops. We saw great use of organic manure. The yields here were comparable with what is achieved elsewhere in the state – in fact, the farmers claimed that the yields were higher. Few farmers in the village were in debt. Here we saw farmers who were not complaining of higher input costs. And Sukho Majri is nowhere near a fully organic-based agriculture.

Certainly, Sukho Majri has a different climate than say Sirsa. But the same *principles* can be applied anywhere.

It is often argued that if we are to feed the millions in the country then we need to step up our yields even higher. There is little doubt about this. But it is often forgotten that the average yield can be increased either by (a) very high yields at one point and low yields elsewhere, or (b) a moderate increase all over.

We have seen the problems with the former – the need to increase inputs vastly in small areas, (and hence extract them excessively, or transfer them from long distances and transfer the output back again), leading to increased costs, and other ecological problems that ultimately impact on the extractions themselves. On the other hand, the latter strategy has the advantage of requiring moderate increases in inputs, meaning not only moderate disruptions in natural cycles – but also decreasing substantially the cash burden on farmers. The output too would be relatively more equitably distributed.

A question can be raised and should be raised: Will this be effective in meeting our food and other needs?

Before we look at this, we would like to emphasise one thing. Whether the approach above can meet our needs or not, one thing is certain – the current approach, as exemplified by Punjab and Haryana, certainly cannot. It is ecologically, economically, financially – and hence socially and politically – unsustainable.

Coming to the alternative approach – there is little doubt that a moderate increase in yields and productivity spread out over large areas can meet our food requirements.

A look at the overall figures will be instructive. In 1997-98, the all-India area under foodgrains was 123.85 m ha, and output was 192.26 m tons. (Rice and wheat accounted for 56% of this). This is equivalent to an average yield of 1552 kg/ha. If we can achieve an increase of 100 kg/ha in this, we will get an additional output of 12 m tons. The area in Punjab under foodgrains in the same year was 5.951 m ha (93% of it under wheat and rice). To get an increase of 12 m tons output, yields would have to increase by 2050 kg/ha. Reasonable increases in yields can be obtained by a variety of measures that will also need only reasonable inputs, with moderate impacts. Extreme increase in yields will need excessive increase in inputs, with large impacts. Of course, this calculation is only indicative of the broad principle, and actual planning would have to take into consideration the differences in land quality, crops other than wheat and rice and so on. But there is little doubt that a decentralised approach can work, and meet our requirements with only moderate impacts.

Indeed, *only such an approach* can meet our needs. And meet it at lesser costs – both capital and recurring, and lesser ecological costs.

More ... How Much is Possible and How Much is Enough

Of course, such a system may not create islands of prosperity and opulence in in midst of poverty. It may not lead to "showcase" agricultural systems of "spectacular" (though unsustainable) performance. It will certainly not help grow sugarcane or rice in deserts.

Take rice – rice is a crop that can and is grown without irrigation in Kerala or Chattisgadh.Yet, in Punjab it cannot grow without huge extractions of groundwater. If we insist on growing rice here (for reasons of higher returns or other reasons), there is no other way but to keep extracting more, and more – till maybe one day there is nothing left. At that point, the catastrophic collapse of the systems would play havoc.

Yet, this is precisely the approach we are following. At a public meeting in Haryana, held to help us interact with the people on these issues, the problems facing Haryana's agriculture, including groundwater depletion were presented. One person got up and said "All our problems are because we do not have enough water. We need more. More water and all our problems will be solved." This, when Punjab is consuming 35 MAF of water and Haryana 27 MAF every year in agriculture.

It is not just an individual, we saw that this belief was widely prevalent. The groundwater levels are falling – so bring in more water and problem solved. Simple. Deceptively simple. Deceptive because it does not answer the question – bring from where? And even if there is an answer today – what, when *that* source reaches its limits? Deceptive because it looks at only the supply side – no thought is given as to why the groundwater levels are falling; and whether the cause for that (rice in Punjab to continue our example) is justified. Deceptive because it hides the fact that unless you put a limit to the cause behind falling groundwater, no matter how much you bring in from outside, it won't be enough…ever.

In other words, unless we also pay attention to how (and how much and for what) the water is being used, we will always need more and more water – which means higher and higher extractions, which will be then justifications for large dams,over-extractions of groundwater and so on – unsustainability.

No system can be sustainable – ecologically, economically, socially – unless it pays attention to this "other side" – namely, use or consumption – and thinks about the limits and ranges for this.

One dimension of this is the justification of the end-use, and whether this end-use is the optimal use of resources. For example, growing rice – on the large scale as is being done today in Punjab – is this justified? Is this the optimal use of water resources? *Only if this end-use is justified can the extractions be justified.* In general, trying to grow crops unsuited to the agro-climatic conditions would be unjustified – or at least, a sub-optimal and high-cost strategy.

The other dimension of this is that any system, whether the end-uses justify it or not, will have its natural limits.[15] This means that we have to live with a recognition that our consumption would eventually have to have some limits. What these limits are, how these will be defined, and what are the implications of this for our production processes are some of the most critical issues that humankind needs to address.

Some people would say that this is an anti-development view, and will quickly pounce on this statement saying – Ah! So you want India to live in the dark ages. You don't want development. But this would be a distortion of what we are saying. We are not against development – or consumption. But we want to emphasise that this will have limits – must have limits. (And that development is not just about increasing consumption). That we will have to make a distinction between needs that are basic and needs that are – for the want of a better word – luxury. We believe that the former can be comfortably met – that sustainable prosperity is possible. But it will be difficult to meet the needs of luxury without crossing the limits of sustainability. Without paying huge costs.

We agree that "More" is a legitimate element of the goals of any development process. But we would say that development is not simply "more and more". And that there are types and types of "more." After all, Oliver Twist also asked for "More". And in recent years, *"Yeh Dil Mange More"* is sought to be projected as the aspiration of the nation[16]. But these two refer to two very different types of needs. This distinction should be recognised as also the fact that there will be limits to the needs that can be met. And that there will be costs in meeting these needs, costs that will escalate sharply as we reach the limits imposed by nature.

The irrigation/agricultural systems in Punjab and Haryana show what can happen as we reach these limits, and the kind of costs – financial, economic, ecological, social – that we have to pay to push these limits. They raise fundamental issues in terms of how much, how and for what to extract from nature. The developments in Punjab and Haryana show the close interdependence between ecological, economic and social sustainability. In this, they exemplify the biggest developmental challenges to India – and also show the possible directions for the country to meet its developmental objectives.

We believe that this is the most important message offered by our study.

(Shripad Dharmadhikary established Manthan Adhyayan Kendra in 2001. Earlier he was a full-time activist of the Narmada Bachao Andolan for 12 years (1988-2001).)

Endnotes

1 BBMB 2002a: Page 9.

2 Dam builders sometimes say that the weir or barrage will cause no submergence, but this is not strictly true as it can affect areas through higher backwater during times of flood.

3 Government of Punjab 2002: Page 104.

4 Even with this productivity, the economic viability of such large dam projects is increasingly being questioned.

5 To the extent that the same combination is used with other means of irrigation, these will add to the costs of those particular means too.

6 Except possibly the contractors.

7 Chapter 26 *Irrigation and Power,* First Five Year Plan.

8 Raj 1960: Page 3.

9 What the words "*of this scope*" imply is noted in the paragraph previous to the one quoted and we reproduce the same here: "In the case of Bhakra Nangal project, several aspects of it would strike one, even at a first glance, as raising issues of considerable importance of economic point of view......The implications of all these clearly deserve to be pursued, and judged alongside the merits of the scheme, in a comprehensive economic appraisal. Once this is done with reference to explicitly stated criteria, and the project ranked in order of preference with other competing projects of a comparable kind, non-economic considerations, as also the economic imponderables, can be introduced and seen in better perspective."

10 We would like to point out that a very important process, involving eminent experts representing all sides of the large dams debate, in the form of the World Commission on Dams has addressed very comprehensively, very convincingly this question. The WCD with 12 members representing dam builders, engineering companies, NGOs, affected peoples movements etc., was set up in 1998 to assess the development effectiveness of large dams worldwide and come out with a set of criteria and guidelines (only) under which large dams should be built. The unanimous report of the WCD was published in November. 2000 and provides a set of core values, strategic priorities, policy principles and guidelines under which new dams should be built. See *www.unep-dams.org.*

11 Shukla, Shrilal 1968: '*Raag Darbari*', Rajkamal Paperbacks, New Delhi : Page 16.

12 Can be translated as: "*The denizens of this co untry are by tradition poets*". They get captivated by a thing before even understanding it and compose poetry on it. Looking at the Bhakra Nangal dam they can say "*Aah! God has once again chosen the land of Bharat (India) to display his miracles*".

13 At least, were in dynamic equilibrium till large scale human interventions disrupted them.

14 Sometimes we are ridiculed as being "*doomsday-ers*".

15 These limits may not be sharply defined points but rather fuzzy boundaries, which would be a function of the costs – financial, ecological, economic, social.

16 For those not familiar – this has been for long the slogan of a Pepsi ad campaign in India. It translates to "*The heart asks for more*".

Index

I

K

M

N

P

R

S